Synthesis Lectures on Chemical Engineering and Biochemical Engineering

This series publishes short books on all aspects of chemical engineering, covering the analysis or design of chemical processes to effectively convert materials into more useful materials or energy. The books will focus on fundamental aspects necessary for chemical engineering design including chemistry, math, physics, and sometimes biology to improve the quality of life by inventing, optimizing, and economizing new technologies and products.

Javid Ahmad Parray · Mohammad Yaseen Mir ·
Nusrat Shafi · A. K. Haghi

Ozone Technology for Food Processing and Preservation

 Springer

Javid Ahmad Parray (iD)
Department of Environmental Science
GDC Eidgah Srinagar
Srinagar, Jammu and Kashmir, India

Nusrat Shafi
Department of Chemistry
GDC Eidgah Srinagar
Srinagar, Jammu and Kashmir, India

Mohammad Yaseen Mir
Centre of Research for Development
University of Kashmir
Srinagar, Jammu and Kashmir, India

A. K. Haghi
Edinburgh, UK

ISSN 2327-6738   ISSN 2327-6746 (electronic)
Synthesis Lectures on Chemical Engineering and Biochemical Engineering
ISBN 978-3-031-81460-0   ISBN 978-3-031-81461-7 (eBook)
https://doi.org/10.1007/978-3-031-81461-7

This Springer imprint is published by the registered company Springer Nature Switzerland AG
The registered company address is: Gewerbestrasse 11, 6330 Cham, Switzerland

If disposing of this product, please recycle the paper.

Contents

Ozone: Physical and Chemical Properties

1

1.1 Introduction

1.1.1 Concept, Structure, and Importance of Ozone

Mustafa (1990) states that photochemical ozone formation occurs in the stratosphere, high-voltage electrical arcs, photochemical smog, ultraviolet (UV) sterilization lamps, and gamma radiation plants. The distinctively clean, fresh air after a rainstorm indicates newly formed ozone in our environment. Ozone may be a more environmentally friendly option for various food applications than conventional methods thanks to the enactment of new rules and broad spectrum application. Ozone has been used in food processing because of its Generally Recognised as Safe (GRAS) status and the US Food and Drug Administration's (FDA) 2001 certification as an antibacterial agent for direct food contact. The remaining chemicals and reaction byproducts have raised questions about chemical sanitizing agents. Trihalomethanes and chloramine compounds, for instance, are byproducts of chlorination that can potentially cause cancer (Pascual et al. 2007). There have been no reports of harmful health effects from ozone reaction products resulting from the oxidation of organic molecules, such as aldehydes, ketones, or carboxylic acids (Pascual et al. 2007). Another option to chlorine for stopping the synthesis of halogenated organic molecules is ozone. However, ozone's physicochemical characteristics determine how effective it is on food and food products within the food sector. The physical, chemical, and antibacterial characteristics of ozone as they relate to the food business are covered in this chapter.

Ozone's structure is what gives it its significant reactivity. Three oxygen atoms make up the ozone molecule. Every oxygen atom has two unpaired electrons in its valence shell, each occupying a 2p orbital. This indicates that three oxygen atoms are joined during its production, with the center oxygen being rearranged from the two 2s and 2p atomic orbitals of the valence band in a plane sp^2. Beltran (2003) reported rearranging

© The Author(s), under exclusive license to Springer Nature Switzerland AG 2025
J. A. Parray et al., *Ozone Technology for Food Processing and Preservation*, Synthesis Lectures on Chemical Engineering and Biochemical Engineering,
https://doi.org/10.1007/978-3-031-81461-7_1

1

the three new sp^2 hybrid orbitals results in a triangle with an oxygen nucleus at its center or an angle of 116° 49'. The configuration of the ozone molecule derives from how the sp^2 and $2p^2$ orbitals are joined, resulting in two 9-molecular orbitals that move across the ozone molecule. Consequently, the four potential structures combine to generate a hybrid, the ozone molecule (Guzel-Seydim et al. 2004; Gunten 2003). The electronic structure of the molecule is responsible for the high reactivity of ozone; in some resonance structures, the absence of electrons in one of the terminal oxygen atoms confirms the electrophilic nature of ozone, whereas the presence of excess negative charge in another oxygen atom confers a nucleophilic nature (Beltran 2003).

1.1.2   Ozone Generation

The general ozone generation was explained by Rice et al. (1981). A diatomic oxygen molecule must first split to produce ozone. As a result, the liberated radical oxygen can combine with another diatomic oxygen to create the triatomic ozone molecule. However, a significant amount of energy is needed to break the O–O bond. Ozone can be produced by using corona discharge techniques and ultraviolet radiation (with a wavelength of 188 nm) to start the production of free radical oxygen. The corona discharge approach typically produces ozone in commercially viable quantities. In a corona discharge, there are two electrodes: the low-tension electrode (ground electrode) and the high-tension electrode. A ceramic dielectric medium separates them, offering a slight discharge gap. Some of these collisions happen when the electrons have enough kinetic energy (about 6–7 eV) to split the oxygen molecule, and each oxygen atom can produce an ozone molecule. One to three percent ozone can be created if air is used as the feed gas via the generator; however, yields can reach as high as six percent ozone when pure oxygen is used (Rice et al., 1981). According to Manley and Niegowski (1967), the ozone concentration cannot be raised above the point at which the creation and destruction rates are equal. Since ozone spontaneously breaks down into oxygen atoms, ozone gas cannot be held (Wickramanayake et al. 1984; Wickramanayaka 1991; Kogelschatz 1988).

1.2   Chemical and Physical Properties of Ozone

C. F. Schonbein, a European scholar, made the initial discovery of ozone in 1839. It was initially applied commercially in Nice in 1907 for municipal water supply purification and in St. Petersburg in 1910. Ozone is the second most potent joint oxidizing agent. Mustafa (1990) states that ozone is created in the stratosphere, photochemical smog, UV sterilizing lamps, high-voltage electric arcs, and gamma radiation plants. Since ozone breaks down quickly at room temperature, it cannot significantly build without ongoing ozone production (Peleg 1976; Miller et al. 1978). Ozone is an almost colorless gas at an average temperature. According to Coke (1993), ozone has a strong, distinct smell, like "fresh air after a thunderstorm." It is easily observable at 0.01–0.05 ppm level (Miller et al. 1978; Mustafa 1990; Mehlman and Borek 1987). In nature, it is present in small amounts. According to Rice (1986), ozone has a longer half-life in gaseous form than in aqueous solution. According to Hill et al. (1982), ozone breaks down quickly into oxygen in clear water and significantly faster in contaminated solutions. According to Rice (1986), ozone is 13 times more soluble in water than oxygen is between 0 and 30 °C. Higher water temperatures speed up the breakdown of ozone (Rice et al. 1981). Ozone is a blue gas at ordinary temperatures, but at concentrations at which it is usually produced, the color is not noticeable. At 112C, ozone condenses to a dark blue liquid. Liquid ozone quickly explodes if ozone-to-oxygen mixtures greater than 20% occur. Explosions may be detonated by electrical sparks or by sudden changes in temperature or pressure. However, in practical use, explosions of ozone are sporadic. The three oxygen atoms in the ozone molecule are arranged at an obtuse angle whereby a central oxygen atom is attached to two equidistant oxygen atoms; the included angle is approximately 116,490, and the bond length is 1.278 A. Four structures of ozone are shown in Fig. 1.1.

Although in low concentrations, ozone is not an extremely toxic gas, at high concentrations, ozone may be fatal to humans (Oehlschlaeger 1978). Ozone is not a very hazardous gas at low doses, but it can be lethal to humans in large amounts. Dogs exposed to ozone (0.65 ppm) for 1–2 h showed fast breathing, while young rats exposed to ozone (0.2 ppm) for 4–6 weeks showed lung distension (Barlett et al. 1974). It was discovered that, depending on the exposure time, ozone concentrations as high as 0.2 ppm can harm

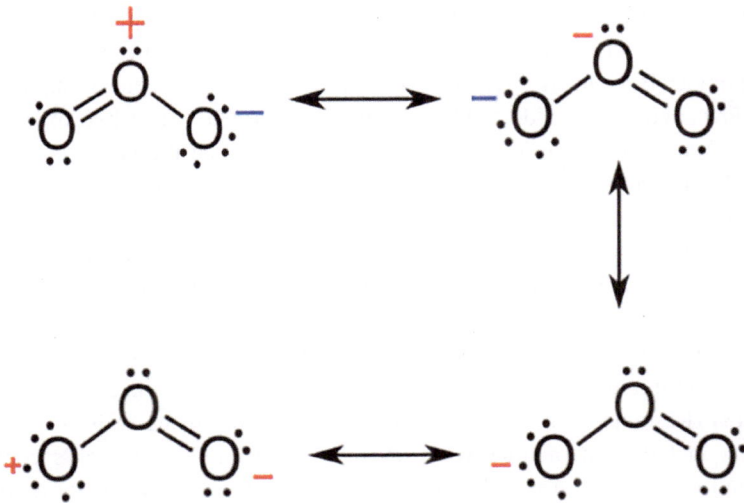

Fig. 1.1 Structure of ozone as a resonance hybrid of the four canonical forms

Table 1.1 Major physical properties of pure ozone

S. no.	Physical properties	Value	References
1	Boiling point	-111.9 ± 0.3 °C	Manley and Niegowski (1967), Guzel-Seydim et al. (2004), Miller et al. (2013)
2	Melting point	-192.57 ± 0.4 °C	Manley and Niegowski (1967), Guzel-Seydim et al. (2004), Miller et al. (2013)
3	Critical temperature	-12.1 °C	Manley and Niegowski (1967), Guzel-Seydim et al. (2004), Miller et al. (2013)
4	Critical pressure	54.6 atm	Manley and Niegowski (1967), Guzel-Seydim et al. (2004), Miller et al. (2013)

the respiratory system to differing degrees (Schwartz et al. 1976). The trachea, bronchi, and alveoli are all affected by damage to the pulmonary system (Schwartz et al. 1976; Castleman et al. 1980). Table 1.1 lists the principal physical characteristics of pure ozone.

1.3   Mechanism of Microbial Inactivation by Ozone

Ozone is recognized as a potent bactericidal and virucidal agent due to its wide range of antibacterial activity in water and wastewater. According to the Surface Water Treatment Rule of 1989, water providers in the United States must administer a disinfectant with a suitable concentration (C) for a sufficient amount of time (t) to eradicate a specific number of germs (Botzenhart et al. 1993). According to Botzenhart et al.'s (1993) analysis of the inactivation of Bacillus subtilus spores, ozone is a more potent sanitizer than chlorine dioxide. Both in the gaseous phase and in solution, ozone is highly unstable, breaking down into superoxide ($.O_2-$), hydroperoxy ($.HO_2$), and hydroxyl (HO) radicals. These free radicals' ability to oxidize is thought to be the cause of ozone's high reactivity (Manousaridis et al. 2005). Numerous species, including spores and vegetative cells, as well as Gram-positive and Gram-negative bacteria, have been subjected to studies on the bactericidal effects of ozone (Ishizaki et al. 1986; Restaino et al. 1995). Ozone is effective against food-related microorganisms in studies involving Gram-positive bacteria (such as Staphylococcus aureus, Listeria monocytogenes, Bacillus cereus, and Enterococcus faecalis), Gram-negative bacteria (such as yeasts Yersinia enterocolitica and Pseudomonas aeruginosa), Candida albicans, Zygosaccharomyces bacilli, and Aspergillus niger spores) (Restaino et al. 1995). Ozone's ability to inactivate cells is a multifaceted process that includes interactions with several components of the cell's content (e.g., enzymes and nucleic acids) and the cell membrane and wall (such as unsaturated fats). Ozone may damage proteins and DNA (Komanapalli and Lau 1996), ozone may harm membrane glycoproteins and glycolipids (Guzel-Seydim et al. 2004), membrane-bound enzymes (Murray et al. 1965), and ozone may cause singlet oxygen contained in cells to oxidize double bonds. According to Victorin (1992), there are two main ways that ozone breaks down target organisms: (1) it oxidizes amino acids, sulfhydryl groups, and enzymes, peptides, and proteins to smaller peptides; and (2) it oxidizes polyunsaturated fatty acids to acid peroxides. Microorganisms become inactive when their cell envelope is disrupted or when they disintegrate and cause cell lysis. The ozone treatment of Bacillus spores appears to inactivate them by breaking down their outer layer, which comprises about 50% of their volume. This exposes the spores' cortex and core to the ozone's action (Foegeding 1985; Khadre et al. 2001). Young and Setlow (2004) found that ozone damages spores' capacity to germinate rather than causing DNA damage to kill the spores. The scientists postulated that spore germination abnormalities result from damage to the spores' inner membrane.

There is disagreement about which is more critical in this inactivation mechanism: molecular ozone or the free radicals generated when it breaks down. Whether radical species or molecular ozone causes the inactivation of microorganisms is a matter of debate. While some researchers (Finch et al. 1992; Labatiuk et al. 1994; Hunt and Marinas 1997) contend that indirect reactions with radicals are the primary mechanism for inactivation, others (Bancroft et al. 1984) suggest direct reaction with molecular ozone. Microorganisms will probably differ in how much of a difference direct and indirect reactions with ozone have in triggering microbial inactivation responses. Three steps of ozone breakdown have been identified: initiation, promotion, and inhibition. The highly reactive hydroxyl radical is formed during the initiation step due to the generation of free radicals such as superoxide radical ions and hydroperoxide radicals. These hydroxyl radicals are one of the things that cause the breakdown of ozone. In the promotion step, promotors such as formic acid, glyoxylic acid, primary alcohols, and aryl groups participate in reactions that regenerate the hydroperoxide and superoxide radicals.

On the other hand, no superoxide radical ion is renewed during the inhibitory stage when hydroxyl radicals are consumed by ions such as bicarbonate, carbonate, tertiary alcohols, and alkyl groups (Staehelin and Hoigné 1985; Khadre et al. 2001). Microbial cells typically contain bicarbonate ions, which may function as radical scavengers and prevent germs from becoming inactive. Furthermore, elements that encourage ozone breakdown within the system may cause ozone to dissipate more quickly, necessitating higher ozone concentrations to reach the required inactivation level (Zuma et al., 2009). The subsequent breakdown or lysis of cell walls (presumably by oxidative destruction) associated with ozone is a faster inactivation process than other disinfectants, which require the disinfecting agent to permeate through the cell membrane to be effective (Pascual et al. 2007). According to Scott and Lesher (1963), ozone can cause specific cells to lyse and cause cell contents to seep into the medium. Therefore, ozone-demanding compounds are created throughout the ozone inactivation process. After lysis, Finch and colleagues (1989) discovered that E. Coli cells required 0.06 mg/L of ozone, and they linked this demand to the second stage of inactivation (Kim and Yousef 2000). Generally speaking, every microbe has an innate sensitivity to ozone regarding the range of microbial action. More sensitive than fungi and yeasts are bacteria. Spores are more ozone-resistant than vegetative cells, while Gram-positive bacteria are more susceptible than Gram-negative species. Due to the mechanism of ozone action, which eliminates the pathogen through cell lysis, the development of resistance to ozone disinfection has not been discovered (Pascual et al. 2007).

1.3.1 Viruses

Generally speaking, viruses can withstand higher ozone concentrations than vegetative bacteria (Rojas-Valencia 2011). Ozone was found by Burleson et al. (1975) to be a viral killer. In their waste-water treatment trial, they discovered that, after 15 s of treatment, ozone treatment effectively inactivated the GDVII, encephalomyocarditis, and vesicular stomatitis viruses (Burleson et al. 1975). According to O'Donnell et al. (2012), viruses with lipid encirclement, or lipid bodies, were more ozone-resistant than viruses without this characteristic. Additionally, he said that a therapeutic dose of 0.4 ppm aqueous ozone might inactivate the hepatitis A virus in 5 s. When ozone (0.37 ppm) is added to drinking water at pH 7 for five minutes at five °C, it can effectively control the Norwalk virus, lowering concentrations by more than three logs in ten seconds (O'Donnell et al. 2012). Ozone's mechanical action on viruses causes the capsid to split into individual pieces, releasing RNA and preventing the virus from adhering to the host pili. Furthermore, the DNA from the head can be released by ozone, which can also randomly kill the fibers in the head, collar, contractile sheath, endplate, and tail.

1.3.2 Fungi

Toxins produced by many fungi have the potential to cause foodborne illnesses. While various fungi can cause food to decay, one genus—including the species Aspergillus flavus and Aspergillus parasiticus—produces aflatoxin, which can sicken people and animals. Ozone can kill fungus by oxidizing mold poisons and causing irreversible cellular damage.

According to Zorlugenç et al. (2008), after dried figs were exposed to gaseous ozone, the amount of aflatoxin B1 in the figs was decreased by 13.8 mg L^{-1} ozone gas at 15 and 30 min. Additionally, deactivating A. flavus and A. parasiticus cells only required 15 min of ozone treatment. They also observed that samples of dried figs artificially contaminated with aflatoxin showed an increase in aflatoxin B1 degradation with increasing ozonation time when treated with gaseous ozone and ozonized water for 30, 60, and 180 min (Öztekin et al. 2006; Zorlugenç et al. 2008).

1.4   Biofilms and Ozone

Few research have looked into how ozone affects microbial biofilm removal or prevention. It's yet unknown how ozone influences bacterial biofilms through specific processes. Gram-negative bacteria typically exhibit greater ozone sensitivity than Gram-positive microbes, as noted by Moore et al. (2000). According to Panebianco et al. (2021), oxidative stress can cause structural losses of the extracellular matrix, lowering the total biomass in preformed biofilm. Conversely, preventive ozone treatment of L. monocytogenes planktonic cells is thought to decrease the bacteria's ability to produce the extracellular polymeric matrix. The primary pathogenic bacterial groups that can form biofilms in the dairy environment and cause spoiling are reported in this section. But it's important to note that pathogenic bacteria like L. monocytogenes and P. fluorescens, as well as spoilage bacteria, can develop mixed biofilms in the dairy environment (Maggio et al. 2021).

1.4.1   Pseudomonas

Bacteria in the genus Pseudomonas can change milk and dairy products in several ways. Pseudomonas species, in particular, are accountable for the odd colors and tastes of food and their unwanted odors (Reichler et al. 2021). Because these microbes are so common, they are typically separated in the dairy environment at various production phases. According to Chiesa et al. (2014), the species P. fluorescens, P. koreensis, P. marginalis, P. Rhodesia, P. fragi, P. putida, P. entomophile, P. mendocina, and P. aeruginosa are more frequently isolated from dairy plants. Thermostable enzymes like lipases and proteases may persist following treatments, producing spoiling in final products, even if these bacteria are susceptible to the heat treatments frequently employed in dairy processing (Zhang et al. 2019). Numerous investigations have shown that Pseudomonas isolates from dairy products, milk, and processing settings can produce biofilms. In this context, a recent study (Rossi et al. 2018) demonstrated the connection between the capacity of P. fluorescens strains associated with dairy products to form biofilms and their ability to produce blue pigment.

Numerous investigations have shown that this species of bacteria is generally vulnerable to ozone exposure, regardless of whether they are grouped in biofilms or adhered to everyday surfaces. Greene et al. (1993) demonstrated that ozonated water treatment (0.5 ppm) with a 10 min exposure time was successful in lowering the loads (~4 Log10) of common psychrotrophic spoilage bacteria on stainless steel surfaces, such as P. fluorescens and Alcaligenes faecalis. They also noted that this technology outperformed commercial chlorinated sanitizers used at high concentrations (100 ppm). Similarly, on stainless steel coupons, Dosti et al. (2005) observed that P. fluorescens (ATCC 948), P. fragi (ATCC 4973), P. putida (ATCC 795), Enterobacter aerogenes (ATCC 35,028), E.

cloacae (ATCC 35,030), and B. licheniformis (ATCC 14,580) were all effectively treated with ozone (0.6 ppm for 10 min). Marino et al. (2024) have revealed the sensitivity of P. fluorescens to ozone treatments, demonstrating the efficacy of ozonated water (0.5 ppm) applied in both static and dynamic settings on biofilms. The scientists also stated that after 60 min of treatment, ozone in gaseous form (20 ppm) was reduced by 5.51 Log CFU/cm^2. In a multilaminated aseptic food packaging material and stainless steel, Khadre and Yousef (2001) investigated the effects of ozone on bacterial biofilms and dried films of B. subtilis spores and P. fluorescens. P. fluorescens in biofilms was more successfully inactivated by ozone on stainless steel than on the multilaminated packing material. Shelobolina et al. (2018) examined the effect of dissolved ozone (2, 5, and 7 ppm for 10 and 20 min) on P. aeruginosa biofilm formed on glass. The ozone effect was examined using a regression equation, which revealed a correlation between biofilm inactivation and concentration and contact time (predicted D-values: 11.1, 5.7, and 2.2 min at 2, 5, and 7 ppm, respectively). The same scientists investigated the effects of dissolved ozone (5 ppm for 20 min) on the inactivation of biofilms on different surfaces. The results showed that those grown on ceramics were more difficult to inactivate than biofilms produced on plastic materials. When combined with other technologies, ozone is effective against Pseudomonas biofilm. For instance, ozone water mixed with a hydrogen peroxide solution worked well against the biofilm of P. fluorescens. Accordingly, synergistic disinfection effects were demonstrated by a sequential treatment using 1.0 and 1.7 mg/L of ozone and 0.8 and 1.1% of hydrogen peroxide (Tachikawa and Yamanaka 2014).

1.4.2   Bacillus

The genus Bacillus is significant among spore-forming bacteria since it contains foodborne pathogens and bacteria that can cause spoiling in milk and dairy products. Bacillus species are widely distributed, rod-shaped, motile, Gram-positive bacteria that are versatile and adaptable to various environmental circumstances. They can also survive the many stages of dairy product manufacture and processing (Shemesh and Ostrov 2020). According to Kumari and Sarkar (2016), B. licheniformis, B. cereus, B. subtilis, B. thuringiensis, B. weihenstephanensis, B. mycoides, B. sporothermodurans, and B. megaterium are the most frequently occurring species in dairy environments. According to Ostrov et al. (2019), Bacillus may cling and remain on various surfaces. They can also form distinct types of biofilms, such as bundles in the liquid phase and pellicles at the air–liquid interface.

Furthermore, Bacillus can produce heat-resistant spores that endure pasteurization procedures. Numerous studies have highlighted the beneficial effects of ozone on biofilms produced by Bacillus linked to dairy products. According to a recent study, B. cereus biofilms grown on polypropylene and stainless steel could be effectively treated with gaseous ozone at 45 ± 2 ppm (Harada and Nascimento 2021). A different study (Babu et al. 2016) examined the impact of ozonated water on B. cereus biofilms cultured on dairy processing membranes and found that treated membranes had an average reduction of 1.0 Log CFU/cm^2. The effectiveness of ozone in conjunction with NaOH, a cleaning agent in situ, was demonstrated on biofilms produced on stainless steel by B. subtilis and B. amyloliquefaciens and compared to NaOH alone, which took 240 s to entirely remove the film from the stainless steel coupons, 1.4 ppm of ozone combined with 1% NaOH resulted in a higher inactivation of biofilms (60 and 120'') (Tiwari et al. 2017).

References

Babu, K.S.; Liu, Z.; Amamcharla, J.K. 0713 Use of fluorescence-based Amaltheys analyzer for studying the effect of pH and heat on whey protein interactions in reconstituted milk protein concentrate. J. Anim. Sci. 2016, 94, 341–342. [CrossRef].

Bancroft K., Chrostowski P., Wright R. L. and Suffet I. H. (1984) Ozonation and oxidation competition values. Wat. Res. 18, 473–478.

Bartlett, J.D., Faulkner, C.S., and Cook, K. 1974. Effects of chronic ozone exposure on lung elasticity in young rats, J. Appl. Physiol. 37:92–96.

Beltran, F.J. (2003). Ozone Reaction Kinetics for Water and Wastewater Systems (1st ed.). CRC Press. https://doi.org/10.1201/9780203509173.

Botzenhart K., Tarcson G. M., Ostruschka M. Inactivation of bacteria and coliphages by ozone and chlorine dioxide in a continuous flow reactor. Water Sci. Tech. 1993; 27:363–370.

Burleson GR, Murray T, Pollard M (1975). Inactivation of viruses and bacteria by ozone, with and without sonication. Applied Microbiology 29(3):340-344.

Castleman WL, Dungworth DL, Schwartz LW, Tyler WS. 1980. Acute respiratory bronchiolitis: An ultrastructural and autoradiographic study of epithelial cell injury and renewal in rhesus monkeys exposed to ozone. Am J Pathol 98:811–840.

Chiesa, F.; Lomonaco, S.; Nucera, D.; Garoglio, D.; Dalmasso, A.; Civera, T. Distribution of Pseudomonas species in a dairy plant affected by occasional blue discoloration. Ital. J. Food Saf. 2014, 3, 1722. [CrossRef] [PubMed].

Coke AL (1993). Mother nature's best remedy: Ozone. Water Conditioning and Purification, (October) pp. 48–51.

Dosti, B.; Guzel-Seydim, Z.; Greene, A.K. Effectiveness of ozone, heat and chlorine for destroying common food spoilage bacteria in synthetic media and biofilms. Int. J. Dairy Technol. 2005, 58, 19–24. [CrossRef].

Finch G. R., Yuen W. C. and Uibel B. J. (1992) Inactivation of Escherichia coil using ozone and ozone-hydrogen peroxide. Environ. Technol. 13, 571–578.

Finch, G.R., Smith, D.W., 1989. Ozone dose–response of Escherichia coli in activated sludge effluent. Water Research 23 (8), 1017–1025.

Foegeding, P.M., 1985. Ozone inactivation of Bacillus and Clostridium spore populations and the importance of the spore coat to resistance. Food Microbiol. 2, 123–134.

Greene, A.K.; Few, B.K.; Serafini, J.C. A Comparison of Ozonation and Chlorination for the Disinfection of Stainless Steel Surfaces. J. Dairy Sci. 1993, 76, 3617–3620.

Gunten, U.V (2003). Ozonation of drinking water: Part I. Oxidation kinetics and product formation. Water research. 37(7): 1443–67. https://doi.org/10.1016/S0043-1354(02)00457-8.

Guzel-Seydim ZB, Greene AK, Seydim AC. Use of ozone in the food industry. *LWT-Food Science and Technology.* 2004;37(4):453–460. https://doi.org/10.1016/j.lwt.2003.10.014.

Harada, A.M.M.; Nascimento, M.S. Effect of dry sanitizing methods on Bacillus cereus biofilm. Braz. J. Microbiol. 2021, 52, 919–926. [CrossRef].

Hill AG and Rice RG (1982). Historical background, properties and applications. In R. G. Rice (Ed.), Ozone treatment of water for cooling application, pp. 1–37. Ann Arbor, MI: Ann Arbor Science Publishers.

Hunt, N.K., Marinas, B.J., 1997. Kinetics of Escherichia coli inactivation with ozone. Water Res. 31, 1355–1362. https://doi.org/10.1016/S0043-1354(96)00394-6.

Ishizaki, K., Shinriki, N., Matsuyama, H., 1986. Inactivation of Bacillus spores by gaseous ozone. J. Appl. Bacteriol. 60, 67–72.

Khadre, M.A., Yousef, A.E. and Kim, J. 2001. Microbiological aspects of ozoneapplications in food: a review, Journal of Food Science, 6: 1242–52.

Khadre, M.A.; Yousef, A.E. Decontamination of a Multilaminated Aseptic Food Packaging Material and Stainless Steel by Ozone. J. Food Saf. 2001, 21, 1–13.

Kim, J. G., & Yousef, A. E. (2000). Inactivation kinetics of foodborne spoilage and pathogenic bacteria by ozone. Journal of Food Science, 65, 521–528.

Komanapalli, I. R., and B. H. S. Lau. 1996. Ozone-induced damage of Escherichia coli K-12. Appl. Microbiol. Biotechnol. 46:610614.

Kumari, S.; Sarkar, P.K. Bacillus cereus hazard and control in industrial dairy processing environment. Food Control 2016, 69, 20–29.

Labatiuk C. W., Belosevic M. and Finch G. R. (1994) Inactivation of Giardia muris using ozone and ozonehydrogen peroxide. Ozone Sci. Eng. 16, 67-78

Maggio, F.; Rossi, C.; Chaves-López, C.; Serio, A.; Valbonetti, L.; Pomilio, F.; Chiavaroli, A.; Paparella, A. Interactions between L. monocytogenes and P. fluorescens in Dual-Species Biofilms under Simulated Dairy Processing Conditions. Foods 2021, 10, 176. [CrossRef] [PubMed].

Manousaridis, G., Nerantzaki, A., Paleologos, E., Tsiotsias, A., Savvaidis, I., and Kontominas, M. (2005). Effect of ozone on microbial, chemical and sensory attributes of shucked mussels. *Food Microbiol.* 22, 1–9. https://doi.org/10.1016/j.fm.2004.06.003.

Marino, M.; Maifreni, M.; Baggio, A.; Innocente, N. Inactivation of Foodborne Bacteria Biofilms by Aqueous and Gaseous Ozone. Front. Microbiol. 2018, 9, 2024.

Mehlman MA and Borek C (1987). Toxicity and biochemical mechanisms of ozone. Environmental Research, 42: 36–53.

Miller, Fátima A., Cristina LM Silva, and Teresa RS Brandao. 2013. "A review on ozone-based treatments for fruit and vegetables preservation." Food Engineering Reviews 5(2): 77-106.

Miller GW, Rice R G, Robson CM, Scullin RL, Kuhn W and Wolf H (1978). An assessment of ozone and chlorine dioxide technologies for treatment of municipal water supplies. US Environmental Protection Agency Report No. EPA-600/2–78–147. Washington, DC: US Government Printing Office

Moore, G.; Griffith, C.; Peters, A. Bactericidal Properties of Ozone and Its Potential Application as a Terminal Disinfectant. J. Food Prot. 2000, 63, 1100–1106. [CrossRef].

Murray, R. G., S. Pamela, and H. E. Elson. 1965. Location of mucopeptide of selection of the cell wall of E. coli and other gram-negative bacteria. Can. J. Microbiol. 11:547–560.

Mustafa MG. Biochemical basis of ozone toxicity. Free Radic Biol Med. 1990;9(3):245-65. https://doi.org/10.1016/0891-5849(90)90035-h. PMID: 2272533.

O'Donnell C, Tiwari BK, Cullen P, Rice RG (2012). Ozone in food processing: John Wiley and Sons.

Oehlschlaeger, H.F. 1978. Reactions of ozone with organic compounds. Pp. 20–37 in R.G. Rice, editor; and J.A. Cotruvo, editor. , eds. Ozone/Chlorine Dioxide Oxidation Products of Organic Materials. Proceedings of a Conference held in Cincinnati, Ohio, November 17–19, 1976. Sponsored by the International Ozone Institute and the U.S. Environmental Protection Agency. Ozone Press International, Cleveland, Ohio. 487 pp.

Ostrov, I.; Sela, N.; Belausov, E.; Steinberg, D.; Shemesh, M. Adaptation of Bacillus species to dairy associated environment facilitates their biofilm forming ability. Food Microbiol. 2019, 82, 316–324.

Öztekin, S., Zorlugenç, B., & Zorlugenç, F. K. (2006). Effects of ozone treatment on microflora of dried figs. Journal of Food Engineering, 75(3), 396–399. https://doi.org/10.1016/j.jfoodeng.2005.04.024

Panebianco, F.; Rubiola, S.; Chiesa, F.; Civera, T.; Di Ciccio, P. Effect of Gaseous Ozone on Listeria monocytogenes Planktonic Cells and Biofilm: An In Vitro Study. Foods 2021, 10, 1484.

Pascual A., Llorca I., Canut A. Use of ozone in food industries for reducing the environmental impact of cleaning and disinfection activities. *Trends Food Sci. Technol.* 2007;18:29–35. https://doi.org/10.1016/j.tifs.2006.07.015.

Peleg M (1976). Review paper: The chemistry of ozone in the treatment of water. Water Research, 10: 361–365.

Reichler, S.; Murphy, S.; Martin, N.; Wiedmann, M. Identification, subtyping, and tracking dairy spoilage-associated Pseudomonas by sequencing the ileS gene. J. Dairy Sci. 2021, 104, 2668–2683. [CrossRef] [PubMed].

Restaino L., Frampton E.W., Hemphill J.B., Palnikar P., Efficacy of ozonated water against various food-related microorganisms. Appl. Environ. Microbiol., 1995, 61, 3471–3475.

Rice, R.G., 1986. Application of ozone in water and wastewater treatment. In: Rice, R.G., Browning, M.J. (Eds.), Analytical Aspects of Ozone Treatment of Water and WasteWater. Syracuse, The institute, New York, pp. 726.

Rice, R. G., Robson, C. M., Miller, G. W., Hill, A. G., 1981. Uses of ozone in drinking water treatment. J. Am. Water Works Ass. 73 (1), 4457.

Rojas-Valencia MN. 2011. Research on ozone application as disinfectant and action mechanisms on wastewater microorganisms. Formatex. 263–271.

Rossi, C.; Serio, A.; Chaves-López, C.; Anniballi, F.; Auricchio, B.; Goffredo, E.; Cenci-Goga, B.T.; Lista, F.; Fillo, S.; Paparella, A. Biofilm formation, pigment production and motility in Pseudomonas spp. isolated from the dairy industry. Food Control 2018, 86, 241–248. [CrossRef].

Schwartz LW, Dungworth DL, Mustafa MG, Tarkington BK, Tyler WS. 1976. Pulmonary responses of rats to ambient levels of ozone: Effects of 7-day intermittent or continuous exposure. Lab Invest 34:565–578.

Scott DBM, Lesher EC. 1963. Effect of Ozone on Survival and Permeability of Escherichia coli. J Bacteriol 85:567-576.

Shelobolina, E.S.; Walker, D.K.; Parker, A.E.; Lust, D.V.; Schultz, J.M.; Dickerman, G.E. Inactivation of Pseudomonas aeruginosa biofilms formed under high shear stress on various hydrophilic and hydrophobic surfaces by a continuous flow of ozonated water. Biofouling 2018, 34, 826–834. [CrossRef] [PubMed].

Shemesh, M.; Ostrov, I. Role of Bacillus species in biofilm persistence and emerging antibiofilm strategies in the dairy industry. J. Sci. Food Agric. 2020, 100, 2327–2336. [CrossRef] [PubMed].

Staehelin, J. and Hoigné, J., Decomposition of ozone in water in organic solutes acting as promoters and inhibitors of radical chain reactions. Environm. Sci. Technol. **19** (1985) 1206–1213.

T. C. Manley and S. J. Niegowski, *Encyclopedia of Chemical Technology*, *2nd ed.*, Wiley, New York, 1967, **14**, pp. 410–432.

Tachikawa, M.; Yamanaka, K. Synergistic disinfection and removal of biofilms by a sequential two-step treatment with ozone followed by hydrogen peroxide. Water Res. 2014, 64, 94–101. [CrossRef] [PubMed].

Tiwari, M.; Scannell, A.; O' Donnell, C. Effect of ozone in combination with cleaning in place reagent (cip) to control biofilms of spore-formers in food process environment. In Biosystems and Food Engineering Research Review 22; Cummins, E.J., Curran, T.P., Eds.; University College Dublin: Dublin, Ireland, 2017.

U. Kogelschatz. Advanced ozone generation. In Process Technologies for Water Treatment (S. Stucki, ed.), p. 87. Plenum Press, New York (1988).

Victorin K. (1992). Review of the genotoxicity of ozone. *Mutation Res.* 277: 221–228.

Wickramanayaka GB (1991). Disinfection and sterilization by ozone. In S. B. Seymour (Ed.), Disinfection, sterilization and preservation (4th Ed.) (pp. 182–190). Malvern, PA: Lea and Febiyer.

Wickramanayake, G. B., A. J. Rubin, and O. J. Sproul. 1984. Inactivation of Naegleria and Giardia cysts in water by ozonation. J. Water Pollut. Control Fed. 56:983–988.

Young SB, Setlow P. Mechanisms of Bacillus subtilis spore resistance to and killing by aqueous ozone. J Appl Microbiol. 2004;96(5):1133-42. https://doi.org/10.1111/j.1365-2672.2004.022 36.x.

Zhang, C.; Bijl, E.; Svensson, B.; Hettinga, K. The Extracellular Protease AprX from Pseudomonas and its Spoilage Potential for UHT Milk: A Review. Compr. Rev. Food Sci. Food Saf. 2019, 18, 834–852.

ZORLUGENÇ, B., ZORLUGENÇ, F.K., OZTEKIN, S. and EVLIYA, I.B., 2008. The influence of gaseous ozone and ozonated water on microbial flora and degradation of aflatoxin B(1) in dried figs. Food and Chemical Toxicology, vol. 46, no. 12, pp. 3593-3597. https://doi.org/10.1016/j.fct. 2008.09.003. PMid:18824207.

Zuma, F., Lin, J., Jonnalagadda, S.B., 2009. Ozone-initiated disinfection kinetics of Escherichia coli in water. J. Environ. Sci. Health A 44 (1), 48–56. https://doi.org/10.1080/10934520802515335.

Application of Ozone Technology in Grain Processing

2.1   Introduction to Ozone Technology

As a potent disinfectant, ozone is widely used in food processing, water treatment, preservation, and many other environmental applications. Due to its benefits over conventional methods of food preservation, ozone as an oxidant has a wide range of potential uses in the food business. When preparing fruits and vegetables, ozone, either in gaseous or liquid form, is frequently applied to inactivate pathogens and spoilage microorganisms (Cullen et al. 2009). In addition to its broad range of microbial inactivation, ozone can destroy mycotoxins and eliminate storage pests. The fact that excess ozone breaks down quickly to produce oxygen and leaves no residue in food is one of its possible benefits. There have been reports of its effectiveness against a variety of microorganisms, such as viruses, bacteria, fungi, protozoa, and bacterial fungal spores (Cullen et al. 2009; Khadre et al. 2001; Restaino et al. 1995). Ozone has been confirmed as Generally Recognised as Safe (GRAS) for use in food processing due to these benefits, which draw the attention of the food industry (Graham 1997).

The proposed employment of ozone in food grain preservation would address the growing concern over using toxic pesticides to kill storage pests. Grain processors are looking for alternatives to control storage pests due to the Montreal Protocol on substances that harm the ozone layer (Fields and White 2002), rising insect resistance, and rising customer demand for grain-free chemicals. Methyl bromide, phosphine, and aluminum phosphide are frequently used fumigants (pesticides) for grain storage. According to the Montreal Protocol, Methyl bromide use is almost completely phased out among these. According to reports, the continuous use of these pesticides interferes with natural agents' biological control systems, causing insect pest outbreaks, the spread of resistance, unfavorable effects on non-target organisms, and problems for the environment and public health (Collins et al. 2005; Islam et al. 2009; Kells et al. 2001; Pimentel et al. 2007, 2009). The

© The Author(s), under exclusive license to Springer Nature Switzerland AG 2025
J. A. Parray et al., *Ozone Technology for Food Processing and Preservation*, Synthesis
Lectures on Chemical Engineering and Biochemical Engineering,
https://doi.org/10.1007/978-3-031-81461-7_2

necessity for the development of selective insect-control alternatives has been brought to light by the growing concern about their negative consequences (Fields and White 2002). Using ozone as a food grain fumigant is feasible while maintaining environmental and financial sustainability.

For instance, research by several authors (Islam et al. 2009; Pereira et al. 2008a, b; Pimentel et al. 2009) demonstrates the feasibility of fumigating maize that has been held for up to six months at 20 °C economically. This study examines the effectiveness of ozone in food grain storage and preservation, the impact of ozonation on product quality, and the state of ozone application in grain processing.

2.2   Role of Ozone Technology in Grain Processing Industries

The plentiful and limitless supply of O_2 serves as a prelude to industrial ozone manufacturing. When high-voltage electric discharge is applied to oxygen molecules, the atoms rearrange, resulting in ozone (O_3). The substance is a bluish gas with significant oxidizing qualities and an overpowering smell. Triatomic oxygen molecules are created when free oxygen radicals combine with diatomic oxygen to produce ozone. Strong O–O bonds are broken to produce the free oxygen radical, which needs much energy. Ozone can be produced through UV light and corona discharge techniques, which start the synthesis of free radical oxygen. Ozone can be created by chemical, thermal, thermonuclear, and electrolytic processes in addition to photochemical (UV radiation) and electric discharge techniques (Kim et al. 1999). The corona discharge method is typically employed in the commercial production of ozone. Two electrodes—high tension and low tension (ground electrode)—produce a small discharge gap, separated by a ceramic dielectric medium (Mahapatra et al. 2005). Electrons with sufficient kinetic energy (about 6–7 eV) to dissociate the oxygen molecule collide, and an ozone molecule can be generated from each oxygen atom (Guzel-Seydim et al. 2004). 1–3% of ozone can be produced when atmospheric air is used as the feed gas in the generator. However, when high-purity oxygen is used, 16% of ozone can be produced. As a result, the ozone concentration cannot be raised above the point at which the rates of creation and destruction balance. Since ozone gas spontaneously breaks into oxygen atoms, ozone must be created on the operating site.

In Europe, ozonation has been used for years to purify drinking water. Ozone has been shown to have several additional commercial uses, such as wastewater treatment, swimming pool and bottled water disinfection, and cooling tower fouling prevention. Although ozone has not been applied extensively in the US food industry, the US Food and Drug Administration approved its use in bottled water in 1982. It classified it as generally recognized as safe (GRAS). In 1997, the US Department of Agriculture allowed ozone use to recondition recycled poultry cooling water. Following a year-long examination of the global ozone database, an expert panel declared in 1997 that ozone was a GRAS material for use as a food sanitizer or disinfectant when applied in compliance with reasonable manufacturing procedures. Ozone is currently permitted as a disinfectant or sanitizer in foods and food processing in the US since the US Food and Drug Administration did not object to the expert panel's conclusions. Because of its oxidation potential, ozone is a potent antibacterial agent. The food business may benefit greatly from using ozone in numerous ways.

Ozone has been proposed for use in the food industry for a variety of purposes, including food surface cleanliness, equipment cleaning in food plants, wastewater reuse, and reducing the biological and chemical oxygen demands (BOD and COD) of food plant waste (Majchrowicz 1998; Dosti 1998). The ability to be used for multiple purposes makes ozone a potential agent. Despite its uncommon use in the food and dairy industries, ozone has been applied in a few specific areas, including the conversion of green tea to black tea (Graham et al. 1969), shellfish cleaning (Anonymous 1972), and disinfection of chicken carcasses and chill water (Yang and Chen 1979; Sheldon and Brown 1986; Chang and Sheldon 1989). This chapter mainly discusses its chemical features, production, antibacterial properties, application on food surfaces, and application on food plant equipment as an alternative sanitizer.

2.3 Advantages of Ozone Technology in Grain Processing

2.3.1 Decontamination of Microbes in Grains

Cereal and cereal products containing ozone are known to possess antimicrobial properties against a range of microorganisms, including Rhizopus, Fusarium, Cladosporium, Aspergillus, Aspergillus, Fusarium, and El-Desouky et al. (De Alencar et al. 2012; El-Desouky et al. 2012; Isikber and Athanassiou 2015; Latifi et al. 2019; Santos et al. 2016; Savi et al. 2014; White et al. 2013). The kind of strain of microorganisms, the state of the cells, the pH of the sample, the temperature, the humidity, and the growth level all affect the antimicrobial ozone activities (El-Desouky et al. 2012). Two mechanisms underlie the ozone-induced inactivation of microbes: first, the oxidation of sulfhydryl and amino acid groups of protein and enzymes in microbes; second, ozone is also thought to act through the oxidation of polyunsaturated fatty acids (Afsah-Hejri et al. 2020; Pandiselvam et al. 2017; Brodowska et al. 2018). Greene et al. (2012) described the process of inactivation of microorganisms by ozone, which entails cell disintegration, followed by the leakage of cell components and subsequent cell lysis. Studies on maize, rice, peanut, and wheat grains subjected to varying ozone concentrations and exposure times have revealed microbial reduction of more than two log cycles (Brito Júnior et al. 2018; Santos et al. 2016). In grains, ozone penetrates kernels to inactivate the microorganisms present. It enters through the diffusion of ozone on the seed coat or through tiny cracks in the grains (White et al. 2013). As the moisture content of grains increases, the half-life of ozone decreases. Half-life is the amount of time needed to reduce the initial concentration of ozone to half. The decay rate constant of ozone increases with increased paddy grain moisture content (Pandiselvam and Thirupathi 2015). It also changes with the bed thickness of grains. Our earlier research (Pandiselvam and Thirupathi 2015; Ravi et al. 2015) showed that a rise in bed thickness causes the ozone reaction rate, which causes the ozone half-life to decrease. To maximize the half-life and efficacy of ozone treatment, grains must be adequately dried to achieve the appropriate moisture content. Table 2.1 presents an overview of the findings on the decontamination of fungi. Rhizopus is the fungal genera least amenable to ozone treatment; White et al. (2013) found that a concentration of 15,000 mg/L of ozone can minimize infections in maize. Penicillium and Aspergillus species differed in their tolerance levels; the former was more accommodating. Under ozone treatment, the morphological features of Aspergillus were examined in peanuts using an optical microscope. Pigment degradation in fungal colonies resulted in a disarray in the fungi's structure, and the oxidation of microbial pigments caused the fungi colonies to turn white (De Alencar et al. 2012). Depending on several factors, various microorganisms have varying degrees of sensitivity to ozone treatment. Therefore, it is essential to select the optimal ozone treatment concentration without compromising the final product's quality.

While the ozone effect might not be able to control the microbial population immediately, the total plate count tends to decrease over the storage period following ozonation. An increase in ozone dosage and exposure time may increase the percentage of microbial reduction. Numerous studies have shown that the moisture content of the grain has an

Table 2.1 Effect of ozone processing on decontamination of fungus

Grain	Fungus species	Conditions of treatment	Efficiency of ozone (%)	References
Corn	*Aspergillus flavus*	Ozone concentration of 47,800 mg/L for 1.8 min	96	McDonough, Campabadal et al. (2011)
Peanut	*Aspergillus flavus A. parasiticus*	21 mg/ L for 96 h	80	De Alencar et al. (2012)
Maize	*Fusarium*	0–500 mg/L for one h	76	White et al. (2013)
Rice	*Aspergillus spp. Penicillium spp.*	10.13 mg/L for 60 h	100	Santos et al. (2016)
Wheat	*Aspergillus spp.*	60 mg/L for 300 min	99.92	Trombete et al. (2017)
Corn	*Aspergillus spp.*	60 mg/L for 480 min	99.74	Porto et al. (2019)
Peanut	*Aspergillus spp.*	1.89 $molO_3 \cdot kg_{peanut}^{-1}$ for 240 min	87	Gomes et al. (2023)

inverse relationship with ozone efficacy because it decreases the half-life of ozone (Pandiselvam and Thirupathi 2015). Additionally, because ozone is denser than air, the ozone in bins diffuses horizontally rather than vertically. Finally, grain bed thickness is essential when ozone diffusion is considered. When bed thickness rises, ozone concentration falls from the bottom of the bin to the grain's surface, with the lowest concentration found there (Pandiselvam et al. 2018). Because of this, the bin needs to have a diffusion channel to guarantee that the ozone concentration is constant throughout the grain mass.

2.3.2 Mycotoxin Degradation

Mycotoxins are naturally occurring substances that are secondary metabolites produced by fungi and present in food. Consuming food or feed tainted with mycotoxin results in mycotoxicoses. The disease can be contracted directly by eating tainted food or indirectly through spore inhalation and skin contact. Mycotoxin may have short-term or long-term effects on human health, depending on its species, kind, and dosage (Afsah-Hejri et al. 2013; Zain 2011). Mycotoxins such as fumonisins (FMN), ochratoxin A (OTA), aflatoxins (AFs), zearalenone (ZEN), deoxynivalenol (DON), citrinin (CTR), and patulin are effectively reduced by ozone (Afsah-Hejri et al. 2020). According to the mechanism of mycotoxin degradation, mycotoxin compounds derived from grains react with ozone to produce low molecular weight products that lessen their toxicity. Ozone concentration, exposure duration, and grain moisture content affect how quickly mycotoxins degrade. Compared to previous treatments, Porto et al. (2019) found that treating maize grits with 60 mg/L of O_3 for 480 min resulted in the most significant reduction in aflatoxin levels. This could be because corn grits have a higher surface area than kernels. So, one of the most critical factors in detoxifying is the moisture level of the grain mass. Qi et al. (2016) confirmed that adding water to humidify maize to a moisture content of 19.6% accelerated the breakdown rates of ZEN and OTA. The production of reactive ions rises with increased moisture content, improving the efficacy of ozone treatment. Chen et al. (2014) discovered similar outcomes with peanuts. Higher half-lives signify that grains have been exposed to enough ozone gas (Alencar et al. 2011; Pandiselvam and Thirupathi 2015; Ravi et al. 2015). Ozone's half-life is temperature-dependent as well. Higher temperatures cause ozone to break down more quickly into oxygen because of its shorter half-life (Miller et al. 2013). Hence, ozone's stability is reduced at higher treatment temperatures. AFB1 degradation rates in maize decreased significantly under the same treatment conditions if the moisture content increased from 13.47 to 20.37% (Luo et al. 2014a). This is in contrast to the findings of Qi et al. (2016), who found that ZEN and OTA degradation increased with increased moisture content. This could be due to variations in materials that may have different surface properties. AFB1 and AFG1 are the aflatoxins most sensitive to ozone, and the most significant reduction in aflatoxin content was observed (Luo et al. 2014b; Porto et al. 2019). The readily attacking C8-C9 double bonds on the terminal furan rings cause the mycotoxin molecules to break into organic acids, aldehydes, and ketones. As a result, AFB2 and AFG2 are less reactive and more resistant to ozonation; of them, AFB2 was the least sensitive to ozone in the food product (Porto et al. 2019). Different toxins have varying sensitivities towards ozone based on the structural arrangement of molecules in mycotoxins. It is also conceivable that the effectiveness of ozone toxicity is contingent on the specific material matrix in which it manifests and is best characterized by seed homogeneity. To produce a product that is both safe and high-quality, mycotoxin levels in grains must be brought down to the appropriate acceptable level using ozone. A more extended exposure period with a lower ozone concentration is preferable to a shorter exposure period with a higher ozone concentration, in addition to the importance of moisture content as a determining factor for ozone therapy. Extending the treatment period to increase detoxification may help expose the grain surface to more ozone.

2.3.3 Management of Insects

Since toxic pesticides such as methyl bromide are phased out and moving towards environment-friendly pesticides, ozone is an alternative fumigant in storage grains with zero residues (Pandiselvam et al. 2019). Vigorous insecticidal activity and ease of generation characterize ozone gas (McDonough et al. 2011). Ozone has been observed to have an adverse effect on both internal and exterior feeders, including Tribolium castaneum, Oryzaephilus surinamensis, Sitophilus granarius, S. zeamais, S. oryzae, and Rhyzopertha dominica (Hansen et al. 2012). The life stage of insects influences ozone toxicity. When exposed to 70 mg/L of ozone for four days, pupae of P. interpunctella are more susceptible than eggs and larvae (Bonjour et al. 2011). Hansen et al. (2013) reported comparable outcomes, obtaining total mortality in all insect life cycle phases except eggs (131 mg/L for eight days). The outer layer of eggs acts as a mechanical barrier to ozone, which could cause this (McDonough et al. 2011).

Therefore, the ozone exposure period for the egg stage could be extended to guarantee the total elimination of the insect pests. The grain amount, duration, and moisture content affected how the insects fumigated with ozone. Three distinct ozone concentrations were measured at two different moisture contents of the paddy, 12.4 and 14.2% (w.b.). In damp settings, it takes longer for an adult R. Dominica to die 100% of the time. Higher moisture causes the ozone migration within grain layers to slow down, requiring longer exposure times to kill insects (Sunisha 2019). The impact of temperature on ozone efficiency was documented by Hansen et al. (2013). At varying doses, insects were subjected to low temperatures of 7.3 and 7.9 °C and high temperatures of 29.6 and 31.6 °C. Insect mortality varies with temperature independently of ozone concentration; however, at higher concentrations, the mortality rate was 100% regardless of temperature (Hansen et al. 2013;

McDonough et al. 2011). The size and shape of the treatment chamber, the characteristics of the grain surface, the type of insect, its life stage, and its exposure (either internally or freely) can all affect the amount of ozone applied to grains. The insects' respiratory system may be the target system, as exposure to ozone can render them fatal. Insects usually reduce oxidative tissue damage by breathing in short bursts after exposure to ozone. This leads to membrane oxidation, DNA strand breaks, and insect pulmonary function changes (Pandiselvam et al. 2019a). Ozone has the ability to regulate both external and internal feeders in grains that have been kept, making it a viable substitute for chemical fumigants. To effectively transport gaseous ozone throughout the grains kept in silos, diffusion ducts or channels must be built.

2.3.4 Starch Modification

The food and bread sectors are the primary users of modified starch. To alter starch's rheological, physicochemical, and thermal characteristics, ozonation is applied to the natural starch (Pandiselvam et al. 2019b). Two main processes are involved in the oxidation of starch molecules. First, a hydroxyl group is converted by oxidation to carboxyl and carbonyl groups, and then starch molecules are depolymerized by cleaving glycosidic links (Pandiselvam et al. 2019b; Goze et al. 2016). Chan et al. (2011) studied the maize starch (dry form) during ozonation of 8 mL/s for 10 min and discovered a decrease in molecular mass (by 22.6%) from native starch. The reduction in chain lengths could be the cause of this. Granular shapes did vary, though, and after being exposed to ozone for one hour, the surfaces of corn starch solution swelled, producing enormous granule sizes (Çatal and Ibanoglu 2012b). Similar results were observed in wheat flour suspensions (Lee et al. 2017), which could weaken starch granules on ozone treatment, and subsequent starch degradation may lead to enhanced water absorption, thereby producing swelling of the starch molecule. As exposure duration increases, hydrolytic groups oxidize to form carbonyl and carboxylic groups, increasing the carbonyl and carboxylic contents of ozonated starch (Goze et al. 2016; Sandhu et al. 2012; Ding et al. 2015). According to Göze et al. (2016) and Sandhu et al. (2012), there was no discernible difference between wheat flour and control samples regarding the impact of ozone on thermal parameters, including gelatinization temperature and enthalpy.

On the other hand, waxy rice starch's thermal characteristics were declining. Polyphenols, proteins, and non-starch polysaccharides that stop starch from oxidizing could cause this variation (Ding et al., 2015; Goz'e et al. 2016). The peak, trough, final, setback, and breakdown viscosity values all decreased, indicating that starch granules weakened due to ozone oxidation. The peak viscosity indicates the water-holding capacity. The pasting parameters were observed to decrease with an increase in ozone duration (Obadi et al. 2018). The viscosity reduction and breakage of starch granules are indicated by breakdown viscosity. Retrogradation of starch is indicated by setback viscosity. Consequently, cooking stability is raised, and the retrogradation tendency of starch molecules may be reduced with a reduction in setback and breakdown viscosities.

2.3.5 Enhancement of Seed Germination

Different doses of ozone application may have an impact on seed germination. In moderation, ozone enhances germination; nevertheless, excessive concentrations have unfavorable consequences (Pandiselvam et al. 2020). The influence of ozone on seed morphological traits, germination ability, and germination energy was discussed. For instance, Alexander et al. (2018) investigated the morphological characteristics of spring wheat seeds that were ozonated for 15 min under various treatment circumstances, such as sprout and total root length. The morphological characteristics of the seedlings underwent notable alterations. Ozone was another factor influencing growing energy. Ozone-treated maize seeds, treated for a maximum of 5 min and 20 days before sowing, showed increased growth energy (Normov et al. 2019). Ozone-treated germinated maize seeds had a seminal root system as long as 0.10 m, double the length of untreated seed roots (Normov et al. 2019). Conversely, seminal root length was not observed in wheat seedlings treated with 60 mg/ L of ozone for 120 or 180 min (Savi et al. 2014). The applied ozone concentration and exposure duration may be related to the negligible changes in root length observed with ozone.

Normov et al. (2019) reported that the germination ability of maize seeds was enhanced to 80% when the initial germination was just 65% and the ozone concentration was between 0.02 and 0.04 mg/Land during 15–25 days before to sowing. This is because seeds' innate energy has been activated. Additionally, higher ozonation damages maize cells and reduces their capacity to germinate. Savi et al. (2014) discovered contradictory results when they exposed wheat to 60 mg/L of ozone for 180 min, resulting in a 12.5% reduction in germination ability (Alexander et al. 2018). Up to 120 min of exposure, there was no discernible impact on seed germination; however, seeds lose some of their ability to germinate when exposed for an additional 180 min. This demonstrates how the amount of ozone and the duration of seed exposure can be limited. Applying ozone to winter wheat seeds promotes germination and increases the seeds' energy during the germination process (Avdeeva et al. 2018). It has been determined that ozone treatments of 14.0–17.0 g s/m^3 for 14 days are the most effective for enhancing seed germination. Ozone technology may help improve germination, but too much ozone inhibits seed germination and growth. There may be a correlation between the type of seed and its morphological features, which could account for the variance in growth potential.

2.4   Conclusion and Future Perspectives

It has been possible to alter the functions of various grain products, such as flour, starch, protein, and grain kernels, by applying ozone gas and solution. The therapy may have a beneficial effect on how wheat grains grind. Excessive strength treatments mainly remove the mycotoxins and associated fungus while oxidizing the chemical components. The process oxidizes lipids, adds carboxyl and carbonyl groups to starch, forms disulfide bonds in proteins, reduces the number of polyphenols, and deactivates endogenous enzymes. The strength of the dough is improved by mild to moderate ozone treatment, although the flour becomes more viscous during the pasting process. Overtreatment causes the dough to become weaker and causes the starch and protein molecules to split. While improving the flour's storage durability, ozone treatment reduces the grains' ability to germinate. The process tends to make different grain products more white. The qualitative characteristics of finished food items are effectively influenced by the impacts of ozone treatment on the characteristics of chemical components in grain. Large air cells and enhanced specific volume are standard features of bread and high-ratio cakes manufactured from flour that has been moderately or mildly processed.

On the other hand, goods manufactured from flour that have undergone excessive treatment typically have lower quality and volume. Ozone-treated flour significantly extended the microbiological shelf life of noodles.

Additionally, ozone treatment boosted the amount of ethanol produced from sorghum flour. Ozone seemed like a "greener" option for bread formation and cake making than potassium bromate and chlorine. Overall, the effects of ozone alterations are determined by the amount and concentration of ozone, the time of the treatment, temperature, moisture content, and the kind of grain (flour versus kernel). Compared to grain in flour form, the chemical components of grain in kernel form are more shielded from ozone oxidation. It is possible to conclude that, in the right circumstances, ozone can work as a "green" agent to provide grain products with the desired functions while greatly enhancing food safety. There are numerous research opportunities to improve ozone use for grain processing. More research and comparison are needed to determine the benefits of treating samples with ozone gas over an aqueous solution. It is still necessary to thoroughly investigate how the grain products' moisture content affects the effects of ozone treatment. Constructing kinetic models to explain the alterations brought about by ozone treatment is still required. The shapes of the grains and the reaction container should be taken into account in the models. It is yet unknown how ozone affects certain chemical elements of grains, including polyphenols, non-gluten grain proteins, and cell wall material. Ozone treatment is still needed for other grains to get the necessary functionality. These grains include oilseeds, pseudocereals, and gluten-free cereals, which are becoming increasingly popular due to consumer demand.

Other food product categories like cookies and pasta—especially the gluten-free varieties—still need to be created using ozone-treated grain flour for improved quality.

Investigating the best formulations to cover up or manage the products' changed flavor and fragrance profile due to lipid oxidation brought on by ozone treatment is essential. It may be possible to lessen the development of an unpleasant aroma in the finished products by properly pre-treating grains and related products (such as defatting) before ozone processing. Selecting and breeding appropriate grain genotypes more resilient to ozone treatment in this regard could be another tactic to replicate the offensive stench. Investigating other quality parameters, particularly the sensory quality, of food products derived from ozone-treated flour would be beneficial. Ozone technology can potentially increase the microbiological shelf life of grain products with medium to high moisture concentrations, such as CSB and fresh noodles. The ozone treatment of grain products can be used with other tactics (such as different food additives) to provide a greater variety of functions for specific uses. It is still necessary to build industrial ozone technology facilities for treating grains and associated items on a big scale, which will need input from other disciplines.

References

Afsah-Hejri, L., Hajeb, P., & Ehsani, R. J. (2020). Application of ozone for degradation of mycotoxins in food: A review. Comprehensive Reviews in Food Science and Food Safety, 19(4), 1777–1808. https://doi.org/10.1111/1541-4337.12594.

Afsah-Hejri, L., Jinap, S., Hajeb, P., Radu, S., & Shakibazadeh, S. H. (2013). A review on mycotoxins in food and feed: Malaysia case study. Comprehensive Reviews in Food Science and Food Safety, 12(6), 629–651. https://doi.org/10.1111/1541-4337.12029.

Alencar, E. R. D., Faroni, L. R., Martins, M. A., Costa, A. R. D., & Cecon, P. R. (2011). Decomposition kinetics of gaseous ozone in peanuts. Engenharia Agrícola, 31(5), 930–939. https://doi.org/10.1590/S0100-69162011000500011.

Alexander, L., Yuri, S., Mikhail, P., Olga, S., Sergey, K., & Irina, L. (2018). Treatment of spring wheat seeds by ozone generated from humid air and dry oxygen. Research in Agricultural Engineering, 64(1), 34–40. https://doi.org/10.17221/106/2016-RAE.

Anonymous. (1972). Use of ozone in sea water for cleansing shellfish. Effluent Water Treatment Journal, 12, 260–262.

Avdeeva, V., Zorina, E., Bezgina, J., & Kolosova, O. (2018). Influence of ozone on germination and germinating energy of winter wheat seeds. Engineering for Rural Development, 17, 543–546. https://doi.org/10.22616/ERDEV2018.17.N128.

Bonjour, E. L., Opit, G. P., Hardin, J., Jones, C. L., Payton, M. E., & Beeby, R. L. (2011). Efficacy of ozone fumigation against the major grain pests in stored wheat. Journal of Economic Entomology, 104, 308–316. https://doi.org/10.1603/ec10200.

Brito Júnior, J. G. D., Faroni, L. R. D. A., Cecon, P. R., Benevenuto, W. C. A. D. N., Benevenuto Júnior, A. A., & Heleno, F. F. (2018). Efficacy of ozone in the microbiological disinfection of maize grains. Brazilian Journal of Food Technology, 21. https://doi.org/10.1590/1981-6723.02217

Brodowska, A. J., Nowak, A., & Smigielski, ´ K. (2018). Ozone in the food industry: Principles of ozone treatment, mechanisms of action, and applications: An overview. Critical Reviews in Food Science and Nutrition, 58(13), 2176–2201. https://doi.org/ https://doi.org/10.1080/10408398.2017.1308313.

Çatal, H., & Ibanoglu, S. (2012b). Structure, Physico-chemical and microbiological properties of ozone-oxidized wheat, corn, potato and rice starches. Journal of Food Science and Engineering, 2(4), 196. https://doi.org/10.17265/2159-5828% 2F2012.04.002

Chan, H. T., Leh, C. P., Bhat, R., Senan, C., Williams, P. A., & Karim, A. A. (2011). Molecular structure, rheological and thermal characteristics of ozone-oxidized starch. Food Chemistry, 126(3), 1019–1024. https://doi.org/10.1016/j.foodchem.2010.11.113.

Chang, Y. H., & Sheldon, B. W. (1989). Application of ozone with physical waste water treatments to recondition poultry process waters. Poultry Science, 68, 1078–1087.

Chen, R., Ma, F., Li, P. W., Zhang, W., Ding, X. X., Zhang, Q. I., & Xu, B. C. (2014). Effect of ozone on aflatoxins detoxification and nutritional quality of peanuts. Food Chemistry, 146, 284–288. https://doi.org/10.1016/j.foodchem.2013.09.059.

Collins, P.J., Daglish, G.J., Pavic, H., Kopittke, R.A., 2005. Response of mixed-age cultures of phosphine-resistant and susceptible strains of lesser grain borer, Rhyzopertha dominica, to phosphine at a range of concentrations and exposure periods. Journal of Stored Products Research 41 (4), 373–385.

Cullen, P.J., Tiwari, B.K., O'Donnell, C.P., Muthukumarappan, K., 2009. Modelling approaches to ozone processing of liquid foods. Trends in Food Science and Technology 20 (3–4), 125–136.

De Alencar, E. R., Faroni, L. R. D. A., Soares, N. D. F. F., da Silva, W. A., & da Silva Carvalho, M. C. (2012). Efficacy of ozone as a fungicidal and detoxifying agent of aflatoxins in peanuts. Journal of the Science of Food and Agriculture, 92(4), 899–905. https://doi.org/10.1002/jsfa.4668.

Ding, W., Wang, Y., Zhang, W., Shi, Y., & Wang, D. (2015). Effect of ozone treatment on physicochemical properties of waxy rice flour and waxy rice starch. International Journal of Food Science and Technology, 50(3), 744–749. https://doi.org/10.1111/ijfs.12691.

Dosti, B. (1998). Effectiveness of ozone, heat and chlorine for destroying common food spoilage bacteria in synthetic media and biofilms. Thesis. Clemson Universtiy, Clemson, SC, pp. 69.

El-Desouky, T. A., Sharoba, A. M. A., El-Desouky, A. I., El-Mansy, H. A., & Naguib, K. (2012). Effect of ozone gas on degradation of aflatoxin B1 and as pergillus flavus fungal. Journal of Environmental & Analytical Toxicology, 2(1), 128. https://doi.org/ https://doi.org/10.4172/2161-0525.1000128.

Fields, P.G., White, N.D.G., 2002. Alternatives to methyl bromide treatments for stored-product and quarantine insects. Annual Review of Entomology 47, 331–359.

Gomes, T., de Souza, H. M., Bortolotto, G. D. S., Escobar, B. A., Furtado, B. G., & Angioletto, E. (2023). Application of ozone in peanut kernels: A multiscale model approach and effects on filamentous fungi decontamination. Journal of Food Engineering, 357, 111649.

Goz´e, P., Rhazi, L., Pauss, A., & Aussenac, T. (2016). Starch characterization after ozone treatment of wheat grains. Journal of Cereal Science, 70, 207–213. https://doi.org/ https://doi.org/10.1016/j.jcs.2016.06.007.

Graham, D.M., 1997. Use of ozone for food processing. Food Technology 51, 121–137.

Graham, H. N., Struder, V. V., & Gurkin, M. (1969). Conversion of green tea using ozone. US patent, 3,484,247.

Greene, A. K., Güzel-Seydim, Z. B., & Seydim, A. C. (2012). Chemical and physical properties of ozone. In C. O'Donnell, B. K. Tiwari, P. J. Cullen, & R. G. Rice (Eds.), Ozone in food processing (1st ed., pp. 19–32). Chichester: Blackwell Publishing Ltd.

Guzel-Seydim, Z., Bever, J.P.I., Greene, A.K., 2004. Efficacy of ozone to reduce bacterial populations in the presence of food components. Food Microbiology 21, 475–479.

Hansen, L. S., Hansen, P., & Jensen, K. M. V. (2013). Effect of gaseous ozone for control of stored product pests at low and high temperature. Journal of Stored Products Research, 54, 59–63. https://doi.org/10.1016/j.jspr.2013.05.003.

Hansen, L. S., Hansen, P., & Jensen, K. V. (2012). Lethal doses of ozone for control of all stages of internal and external feeders in stored products. Pest Management Science, 68, 1311–1316. https://doi.org/10.1002/ps.3304.

Isikber, A. A., & Athanassiou, C. G. (2015). Ozone gas is used to control insects and microorganisms in stored products. Journal of Stored Products Research, 64, 139–145. https://doi.org/10.1016/j. jspr.2014.06.006.

Islam, M.S., Hasan, M.M., Xiong, W., Zhang, S.C., Lei, C.L., 2009. Fumigant and repellent activities of essential oil from Coriandrum sativum (L.) (Apiaceae) against red flour beetle Tribolium castaneum (Herbst) (Coleoptera: Tenebrionidae). Journal of Pest Science 82 (2), 171–177

Kells, S.A., Mason, L.J., Maier, D.E., Woloshuk, C.P., 2001. Efficacy and fumigation characteristics of ozone in stored maize. Journal of Stored Products Research 37 (4), 371–382.

Khadre, M.A., Yuosef, A.E., Kim, J., 2001. Microbiological aspects of ozone applications in food: a review. Journal of Food Science 6, 1242–1252.

Kim, J.G., Yousef, A.E., Dave, S., 1999. Application of ozone for enhancing foods' microbiological safety and quality: a review. Journal of Food Protection 62, 1071–1087

Latifi, Z., Arvanaghi, M., & Daneshniya, M. (2019, December). Effect of ozonation on microbial properties of rice flour. In Paper presented on XIII International Conference on Engineering and Technology (ICETCONF), Osio- Norway.

Lee, M. J., Kim, M. J., Kwak, H. S., Lim, S. T., & Kim, S. S. (2017). Effects of ozone treatment on physicochemical properties of Korean wheat flour. Food Science and Biotechnology, 26(2), 435–440. https://doi.org/10.1007/s10068-017-0059-5.

Luo, X., Wang, R., Wang, L., Li, Y., Bian, Y., & Chen, Z. (2014a). Effect of ozone treatment on aflatoxin B1 and safety evaluation of ozonized corn. Food Control, 37, 171–176. https://doi.org/ 10.1016/j.foodcont.2013.09.043.

Luo, X., Wang, R., Wang, L., Li, Y., Wang, Y., & Chen, Z. (2014b). Detoxification of aflatoxin in cornflour by ozone. Journal of the Science of Food and Agriculture, 94(11), 2253–2258. https:// doi.org/10.1002/jsfa.6550.

Mahapatra, A.K., Muthukumarappan, K., Julson, J.L., 2005. Applications of ozone bacteriocins and irradiation in food processing: a review. Critical Reviews in Food Science and Nutrition 45, 447–461.

Majchrowicz, A. (1998). Food safety technology: a potential role for ozone? Agricultural Outlook, Economic Research Service/USDA, pp. 13–15.

McDonough, Marissa X., Linda J. Mason, and Charles P. Woloshuk. 2011. "Susceptibility of stored product insects to high concentrations of ozone at different exposure intervals." Journal of Stored Products Research 47(4) : 306–310.

Miller, F. A., Silva, C. L., & Brandao, ~ T. R. (2013). A review on ozone-based treatments for fruit and vegetables preservation. Food Engineering Reviews, 5(2), 77–106. https://doi.org/10.1007/ s12393-013-9064-5.

Normov, D., Chesniuk, E., Shevchenko, A., Normova, T., Goldman, R., Pozhidaev, D., & Trdan, S. (2019). Does ozone treatment of maize seeds influence their germination and growth energy? Acta Agriculturae Slovenica, 114(2), 251–258. https://doi.org/ https://doi.org/10.14720/ aas.2019.114.2.10.

Obadi, M., Zhu, K. X., Peng, W., Sulieman, A. A., Mahdi, A. A., Mohammed, K., & Zhou, H. M. (2018). Shelf life characteristics of bread produced from ozonated wheat flour. Journal of Texture Studies, 49(5), 492–502. https://doi.org/10.1111/jtxs.12309.

Pandiselvam, R., & Thirupathi, V. (2015). Reaction kinetics of ozone gas in green gram (Vigna radiate). Ozone: Science & Engineering, 37(4), 309–315. https://doi.org/ https://doi.org/10.1080/019 19512.2014.984158.

Pandiselvam, R., Manikantan, M. R., Divya, V., Ashokkumar, C., Kaavya, R., Kothakota, A., & Ramesh, S. V. (2019b). Ozone: An advanced oxidation technology for starch modification. Ozone: Science & Engineering, 41(6), 491–507. https://doi.org/ https://doi.org/10.1080/019 19512.2019.1577128.

Pandiselvam, R., Mayookha, V. P., Kothakota, A., Sharmila, L., Ramesh, S. V., Bharathi, C. P., & Srikanth, V. (2020). Impact of ozone treatment on seed germination–A systematic review. Ozone: Science & Engineering, 42(4), 331–346. https://doi.org/10.1080/01919512.2019.1673697

Pandiselvam, R., Sunoj, S., Manikantan, M. R., Kothakota, A., & Hebbar, K. B. (2017). Application and kinetics of ozone in food preservation. Ozone: Science & Engineering, 39(2), 115–126. https://doi.org/10.1080/01919512.2016.1268947.

Pandiselvam, R., Thirupathi, V., Chandrasekar, V., Kothakota, A., & Anandakumar, S. (2018). Numerical simulation and validation of ozone gas mass transfer process in rice grain bulks. Ozone: Science & Engineering, 40(3), 191–197. https://doi.org/10.1080/01919512.2017.140 4902.

Pandiselvam, R., Thirupathi, V., Mohan, S., Vennila, P., Uma, D., Shahir, S., & Anandakumar, S. (2019a). Gaseous ozone: A potent pest management strategy to control Callosobruchusmaculatus (Coleoptera: Bruchidae) infesting green gram. Journal of Applied Entomology, 143(4), 451–459. https://doi.org/10.1111/jen.12618.

Pereira, A.d.M., Faroni, L.R.D., Silva Jr., A.G.d.S., Sousa, A.H.d., Paes, J.L., 2008a. Economical viability of ozone use as fumigant of stored corn grains. Engenharia Na Agricultura 16 (2), 144–154

Pereira, A.D.M., Faroni, L.R.D.A., De Sousa, A.H., Urruchi, W.I., Paes, J.L., 2008b. Influence of the grain temperature on the ozone toxicity to Tribolium castaneum. Revista Brasileira de Engenharia Agrı´cola e Ambiental 12 (5), 493–497.

Pimentel, M.A.G., Faroni, L.R.D., Guedes, R.N.C., Sousa, A.H., Totola, M.R., 2009. Phosphine resistance in Brazilian populations of Sitophilus zeamais Motschulsky (Coleoptera: Curculionidae). Journal of Stored Products Research 45 (1), 71–74.

Pimentel, M.A.G., Faroni, L.R.D., TA¯ 3 tola, M.R., Guedes, R.N.C., 2007. Phosphine resistance, respiration rate and fitness consequences in stored-product insects. Pest Management Science 63 (9), 876–881.

Porto, Y. D., Trombete, F. M., Freitas-Silva, O., De Castro, I. M., Direito, G. M., & Ascheri, J. L. R. (2019). Gaseous ozonation to reduce aflatoxins levels and microbial contamination in corn grits. Microorganisms, 7(8), 220. https://doi.org/10.3390/2Fmicroorganisms7080220.

Qi, L., Li, Y., Luo, X., Wang, R., Zheng, R., Wang, L., & Chen, Z. (2016). Detoxification of zearalenone and ochratoxin A by ozone and quality evaluation of ozonised corn. Food Additives & Contaminants: Part A, 33(11), 1700–1710. https://doi.org/10.1080/19440049.2016.1232863.

Ravi, P., Venkatachalam, T., & Rajamani, M. (2015). Decay rate kinetics of ozone gas in rice grains. Ozone: Science & Engineering, 37(5), 450–455. https://doi.org/10.1080/01919512.2015. 1040912.

Restaino, L., Frampton, E., Hemphill, J., Palnikar, P., 1995. Efficacy of ozonated water against various food-related microorganisms. Applied Environmental Microbiology 61, 3471–3475.

Sandhu, H. P., Manthey, F. A., & Simsek, S. (2012). Ozone gas affects the physical and chemical properties of wheat (Triticumaestivum L.) starch. Carbohydrate Polymers, 87(2), 1261–1268. https://doi.org/10.1016/j.carbpol.2011.09.003.

Santos, R. R., Faroni, L. R., Cecon, P. R., Ferreira, A. P., & Pereira, O. L. (2016). Ozone as fungicide in rice grains. Revista Brasileira de Engenharia Agrícola e Ambiental, 20(3), 230–235. https:// doi.org/10.1590/1807-1929/agriambi.v20n3p230-235.

Savi, G. D., Piacentini, K. C., & Scussel, V. M. (2014). Ozone treatment efficiency in Aspergillus and Penicillium growth inhibition and mycotoxin degradation of stored wheat grains (Triticumaestivum L.). Journal of Food Processing and Preservation, 39(6), 940–948. https://doi.org/10.1111/jfpp.12307.

Sheldon, B. W., & Brown, A. L. (1986). Efficacy of ozone as a disinfectant for poultry carcasses and chill water. Journal of Food Science, 51(2), 305–309.

Sunisha, K. (2019). Ozone fumigation in stored paddy: Changes in moisture content upon storage. Journal of Entomology and Zoology Studies, 7(3), 1137–1140.

Trombete, F. M., Porto, Y. D., Freitas-Silva, O., Pereira, R. V., Direito, G. M., Saldanha, T., & Fraga, M. E. (2017). Efficacy of ozone treatment on mycotoxins and fungal reduction in artificially contaminated soft wheat grains. Journal of Food Processing and Preservation, 41(3), Article e12927. https://doi.org/10.1111/jfpp.12927.

White, S. D., Murphy, P. T., Leandro, L. F., Bern, C. J., Beattie, S. E., & van Leeuwen, J. H. (2013). Mycoflora of high-moisture maize treated with ozone. Journal of Stored Products Research, 55, 84–89. https://doi.org/10.1016/j.jspr.2013.08.006.

Yang, P. P. W., & Chen, T. C. (1979). Effects of ozone treatment on microflora of poultry meat. Journal of Food Processing and Preservation, 3, 177–185.

Zain, M. E. (2011). Impact of mycotoxins on humans and animals. Journal of Saudi Chemical Society, 15(2), 129–144. https://doi.org/10.1016/j.jscs.2010.06.006.

Ozone Applications for Fruits and Vegetables 3

3.1   Introduction

These days, vegetables account for a sizeable portion of the food market and a significant portion of the daily diet. In fact, because of their high nutritional content, they are essential for a balanced diet reduced in fat, sugar, and sodium. In addition, vegetables are a great source of non-nutrient compounds such as plant sterols, flavanols, anthocyanins, phenolic acids, vitamins, minerals, dietary fibers, and complex carbs. The World Health Organisation (WHO) recommends that everyone consume at least 400 g of fruits and vegetables daily to boost overall health since eating a wide variety of vegetables helps to provide an appropriate intake of essential nutrients (WHO 2018). By preventing weight gain and lowering the risk of obesity, this consumption reduces the risk of several nonchronic diseases, such as several forms of cancer and cardiovascular disorders (Hartley et al. 2013). In addition to their appealing sensory attributes, vegetables are valued for their flavor, scent, texture, color, gloss, shape, size, and lack of flaws or rotting. Eighty percent of buyers give these products' looks a lot of consideration. Hillman et al. (2003) state that the primary selection criterion seems qualitative.

Nevertheless, several foodborne illness outbreaks have been linked to their intake due to their short shelf life (Denis et al. 2016). This highlights how crucial it is to use tailored treatments to disinfect vegetables and effectively prevent microorganisms' growth. Typical chemical treatments, also known as antimicrobial solutions, are used to improve the shelf life of vegetables. These treatments include hydrogen peroxide, peracetic acid, chlorine, and electrolyzed water. In recent decades, the food industry has most frequently employed the first one—sodium hypochlorite, or chlorine—in aqueous formulations for various purposes (such as washing and spraying). Most scientific works that seek to identify an adequate substitute for chlorinated water also use this treatment as a point of reference. Research has shown that chlorine efficiently treats foodborne pathogens (Gu et al. 2020)

© The Author(s), under exclusive license to Springer Nature Switzerland AG 2025   27
J. A. Parray et al., *Ozone Technology for Food Processing and Preservation*, Synthesis
Lectures on Chemical Engineering and Biochemical Engineering,
https://doi.org/10.1007/978-3-031-81461-7_3

and preserves the product's quality throughout its shelf life (Garcia et al. 2003; Baur et al. 2004a, b). But as people's awareness of food safety and health has grown, so has their disapproval of artificial additives (Ma et al. 2017). Because chlorine reacts with organic matter, bromide, and iodide to form hazardous chemicals in wastewater, including halo acetic acids, trihalomethanes, monochloramine, and organochlorinated byproducts, some European countries have outlawed the use of chlorine (Yang et al. 2014; Shen et al. 2016). According to Glassmeyer (2005), these byproducts are mutagenic, cytotoxic to mammalian cells, and persistent in the environment. They also induce DNA damage. The development and deployment of more green technologies for preserving vegetable safety and quality have always been industry issues, partly because of these limitations and the growing demand for natural additives. Ozone utilization shows promise in this setting, and with current technologies, the vegetable business is becoming more interested in it.

Ozone (O_3) is a potent disinfectant that might satisfy producers' demands, regulatory bodies' approval, and consumer acceptability. In the USA, Ozone was approved as a general recognition of safety (GRAS) in 1995 for disinfecting bottled water. Ozone has been classified as GRAS for direct food contact since 1997. The US Food and Drug Administration (FDA) authorized Ozone in gas and aqueous phases in June 2001 as an antibacterial agent in direct food contact. This was done in response to a petition on food additives from the Electric Power Research Institute (EPRI). After its initial water purification application, Ozone was used in food processing in the European Union in the early 1900s. The European Council of Ministers has approved a proposal that allows natural mineral water to be treated with Ozone.

In France, the French Food Safety Authority (AFSSA, now known as ANSES) issued two judgments in 2003 and 2004 about the safety of treating wheat grains with Ozone as an auxiliary technique before grinding. Based on ozone treatment in a closed sequential batch reactor, the regulatory body has approved using Ozone since 2006 as a processing aid for improving flour quality. In 2019, ANSES issued a ruling supporting the expansion of Ozone in water as a technological tool for cleaning salads that are ready to eat. Subsequently, studies and industrial uses have been carried out to verify that Ozone can replace conventional sanitizing chemicals and offer advantages for safe products with longer shelf lives.

Furthermore, The potential applications of Ozone and its advantages or disadvantages have been thoroughly studied for various plants. Ozone can be used in two different ways while processing vegetables. The harvesting process involves periodically or continually adding gaseous Ozone to the harvested product's storage environment. Aqueous Ozone is introduced during the washing process or right after the harvest of vegetables. In the latter scenario, the product can be sprayed, rinsed, or dipped in water that has dissolved Ozone to wash it. A review of ozone application in the vegetable industry has already been done (Karaca and Velioglu 2007). However, this study aims to compile and synthesize the most recent research on some globally produced and consumed veggies that other works have not addressed. Three types of vegetables have been selected: a fruiting vegetable (tomato), a green leafy vegetable (Lettuce), and a root vegetable (carrot). These three vegetables have been selected for several reasons: They are consumed extensively globally, and their varying growth environments have given them unique qualities. For these reasons, we shall address how Ozone affects these veggies' physical and chemical characteristics and their microbiological, sensory, and nutritional qualities. The unique aspect of the suggested method is that it addresses the veggies' total quality (microbiological, physical, chemical, and nutritional) both during and after washing and storage.

3.2   Use of Ozone Treatment in the Vegetable and Fruit Industry

Vegetables and fruits are highly fragile and perishable foods that can be damaged mechanically, lose water content or deteriorate physiologically. It has been assessed how ozone exposure affects the physiology and quality of several food products. Despite having a considerable oxidizing activity, treating fresh-cut Lettuce with ozonated water did not boost its respiratory activity (Forney et al. 2007). According to Zhang et al. (2005), ozone treatment reduced the respiration rate of freshly chopped celery. Three weeks of exposure to 0.3 ppm ozone did not affect peach respiration or ethylene production. Eliminating ethylene from storage areas or containers can delay fruit deterioration and increase fruit shelf life after harvest. According to Skog and Chu (2001), Ozone can lower the airborne ethylene concentration in a cold storage room, allowing commodities that produce and are sensitive to ethylene to be shipped or kept together for extended periods. After five weeks of storage at 5C and 90% relative humidity in peaches, continuous ozone exposure at 0.3 ppm increased water loss; however, this effect did not occur in grapes after four weeks of storage (Palou et al. 2002). Strawberries were kept for three days at 2 in the air with or without 1.5 ppm ozone by Nadas et al. (2003), then moved to room temperature. After being kept in cold storage, fruits treated with Ozone lost less weight than samples that were not. However, changes were not statistically significant under ambient settings by the conclusion of the following period. The ozone treatment likely decreased the fruit's transpiration loss, but this effect vanished when the fruit was exposed to normal air again. After that, the weight reduction accelerated (Nadas et al. 2003). According to Zhang et al. (2005), there was no discernible change in the overall sugar concentration of celery after ozonated water treatment. Strawberry storage was done by Perez et al. (1999) for three days at 2 in an environment with 0.35 ppm ozone and then for four days at 20 °C. As fruits were stored, sucrose in both treated and untreated fruits dropped. There was a change in the amounts of fructose and glucose from day 0 to day 5. How sucrose is converted into glucose and fructose varies markedly between samples that have been treated and those that have not. Low fructose, glucose, and sucrose levels were detected on the third day of storage. This might result from sucrose breakdown mechanisms activating in response to ozone-induced oxidative stress (Perez et al. 1999).

Fruits and vegetables undergo enzymatic browning due to the activity of a class of enzymes termed polyphenol oxidases, which are present in all plants. According to Zhang et al. (2005), ozone treatment inhibits fresh-cut celery's polyphenol oxidase activity. Additionally, Beltran et al. (2005) found that ozonated water prevented browning during storage while preserving the fresh-cut Lettuce's original appearance.

There have been conflicting reports about how Ozone affects peroxidase activity. Peroxidase activity in blackberries stored at 0.3 ppm ozone decreased gradually over time and stayed much lower than the control. On the other hand, 0.1 ppm of ozone exposure increased peroxidase activity compared to the control group (Barth et al. 1995). Different washing procedures, such as washing with ozonated water, had less impact on polyphenol oxidase and peroxidase activities, making them less appropriate physiological indicators of stress reactions brought on by alternate processing (Baur et al. 2004a, b). Foods contain essential antioxidant molecules such as flavonoids, polyphenols, and vitamins A and C. They might act as oxidative enzymes' natural substrates, like polyphenol oxidase. Ozone's potent oxidizing action is anticipated to result in the loss of antioxidant components. Beltran et al. (2005) observed no significant impact of ozone-washing treatment on the final phenolic content of fresh-cut iceberg lettuce. According to Baur et al. (2004a, b), washing Lettuce with ozonated water caused a modest decrease in caffeic acid derivatives.

Furthermore, Ozone was found to generate pterostilbene and resveratrol phytoalexins in table grapes, increasing the fruit's resistance to diseases in the future (Sarig et al. 1996). Some fruits and vegetables contain vitamin C (ascorbic acid), which adds value to these foods because of its significant nutritional significance. Lewis et al. (1996) reported that Ozone lowers the ascorbic acid content of broccoli florets. Conversely, Zhang et al. (2005) found no statistically significant variation in the vitamin C levels of celery samples treated with ozonated water compared to those that were not. Vitamin C retention following storage was comparable in samples of Lettuce that had been chlorine-washed and ozonated water. According to Beltran et al. (2005), slight drops in vitamin C concentrations have been reported from the initial values. Furthermore, it has been shown that exposure to Ozone causes a rise in the amounts of ascorbic acid in spinach (Luwe et al. 1993), pumpkin leaves (Ranieri et al. 1996), and strawberries (Perez et al. 1999). An antioxidative system that encourages the manufacture of vitamin C from the products' carbohydrate reserves has been linked to changes in ozone-treated goods' vitamin C and sugar contents (Perez et al. 1999).

Blackberries' and strawberries' anthocyanin concentrations were found to be slightly impacted by ozone treatments (Barth et al. 1995; Perez et al. 1999). The anthocyanin content of blackberries was kept in the air, and 0.1 parts per million of Ozone stayed constant. It varied in samples treated with 0.3 parts per million of Ozone during storage. The berry fruit displayed the best red color in samples treated with 0.3 ppm ozone during storage (Barth et al. 1995). Additionally, it was noted that samples treated with Ozone showed a noticeably less noticeable unfavorable color shift in broccoli from green to yellow (Skog and Chu 2001).

Nonetheless, it has been noted that Ozone can alter the surface color of some goods, such as carrots (Liew and Prange 1994a, b) and peaches (Badiani et al. 1996). When peaches, grapes, and citrus fruits were treated with an ozonated atmosphere, no phytotoxic damage to the fruit tissues was seen (Palou et al. 2001). The final product's sensory quality was not negatively impacted by treating celery and carrots with ozonated water (Oztekin et al. 2005). The loss of scent in fruits was the most noticeable impact of Ozone on their sensory qualities. Strawberries in ozone-enriched cold storage experienced reversible fruit fragrance reductions (Perez et al. 1999). Treating strawberries with Ozone has lowered their volatile ester emissions by 40%. There were no discernible variations in the enzyme activity of the treated and untreated samples, ruling out the possibility that this was caused by the aroma biosynthesis-related enzymes lipoxygenase, hydroperoxide lyase, and alcohol acyltransferase (Perez et al. 1999). Nadas et al. (2003) suggested that the fruit's volatile component oxidation was the cause of the decreased fruit scent. The research above demonstrates how different food compositions, ozone dosages, application methods, and timing can affect the physiology and quality of fruits and vegetables when exposed to Ozone. Under the right circumstances, Ozone can minimize the physiological harm that fruits and vegetables sustain while preventing microbial deterioration and some illnesses. Trials involving the application of Ozone to various items are necessary to identify these requirements precisely.

3.3 Factor Affecting Ozone-Processing Efficiency

Internal and external elements can influence Ozone's effectiveness, and it can be challenging to forecast how Ozone will behave in fruits and vegetables when certain substances like organic matter and environmental variables are present. Furthermore, Table 3.1 from O'Donnell et al. (2012) lists parameters and factors that affect the effectiveness of ozone treatment, and it concludes with bibliographical research.

Table 3.1 Extrinsic and intrinsic factors influencing the efficacy of Ozone (O'Donnell et al. 2012)

	Parameters	Factors
Extrinsic factors	Water quality	pH, organic matter, pressure, and temperature
	Air quality	Air relative humidity
	Ozone treatment	Concentration and treatment time application method
Intrinsic factors	Food product	Type of fruit and vegetable, weight, characteristics of the product surface, and surface area
	Microbial load	The activity of water (aw) Characteristics of microbial strains, the physical state of bacterial strains, natural microflora, artificially inoculated microorganisms, and population size

3.3.1   Extrinsic Parameters

Ozone's effectiveness as a disinfectant depends on several environmental factors, including medium pH, temperature, humidity, and the amount of organic water surrounding the product and microorganisms, in addition to the amount used (Restaino et al. 1995). Ozone is unstable in both aqueous solutions and the air. The breakdown rate of Ozone in the treatment environment needs to be as low as feasible to guarantee a high degree of microbiological killing by Ozone. The ozone decomposition reaction is influenced by pH. Ozone was shown to be reasonably stable and decompose relatively moderately in acidic settings (pH about 3.0–4.0) (Gurol and Singer 1982; Pan et al. 1984). The primary mechanism of ozone breakdown, hydroxyl radical production, causes ozone degradation to accelerate with increasing pH (although always at or around 7.0). The significance of the hydroxide ion initiation step and peroxy-radicals increases ozone decomposition in alkaline settings (pH of 9.0 and above) (Buffle et al. 2006). As a result, it was shown that microbial survival is better at a pH greater than or equal to 7.0 and that Ozone kills microorganisms considerably more quickly at lower pH values. According to Patil et al. (2010), the time needed for ozone treatment to reduce two strains of Escherichia coli in apple juice by 5 log CFU (Colony-Forming Unit) mL^{-1} was faster at pH 3.0 (4 min) than at pH 5.0 (18 min).

Furthermore, E. Coli had a better chance of surviving at pH 8.0 across various water types than at lower pH values (Jamil et al. 2017). In contrast to lower pH, the ideal buffer pH for Staphylococcus aureus survival is 5.5–6.0 (Britton et al. 2020). It is also crucial to consider the quantity of organic matter because pH significantly impacts the percentage of disinfection by hydroxyl radical formation triggered by the ozone chain reaction. Organic or inorganic suspended particles may contribute to an increased need for Ozone. It is well known that an organic load present during treatment reduces the efficiency of Ozone in terms of inactivating bacteria through ozone consumption (Cho et al. 2003). The rate at which humic acid inhibited the inactivation of E. coli by Ozone was lower in the Hunt and Mariñas (1999) investigation than when natural organic matter was absent. This was caused by the dissolved Ozone breaking down more quickly in the presence of organic materials, reducing the time E. Coli cells were exposed to this disinfectant.

Furthermore, a model orange juice solution demonstrated a quicker rate of E. coli inactivation (1 min) than a low-powder juice (6 min).

In contrast, unfiltered juice showed inactivation after 15 to 18 min (Patil et al. 2009). These findings indicated that both gaseous and aqueous Ozone's antibacterial properties are considerably hampered by organic matter. Furthermore, Restaino et al. (1995) discovered that this organic matter significantly impacts ozone efficiency more than the quantity of organic suspended materials.

According to their findings, adding 20 ppm of soluble starch did not significantly alter the death rates of tested bacteria (S. aureus, Listeria monocytogenes, E. coli, and Salmonella typhimurium) in ozonated water containing organic material, but adding 20 ppm of bovine serum albumin did significantly lower those rates (Restaino et al. 1995). Ozone's stability, reactivity, and solubility are all influenced by factors that may also impact the gas's effectiveness. The effectiveness of Ozone as a biocide is influenced by temperature. According to O'Donnell et al. (2012), lowering an aqueous medium's temperature enhances Ozone's solubility and stability, increasing its availability and, ultimately, its efficacy. Ozone's ability to inactivate is connected with a drop in temperature. Ozone loses stability and solubility as temperature rises, and its breakdown rate accelerates (Rice et al. 1981). Ozone concentration and treatment duration are two external factors that affect ozone efficiency regardless of the type of Ozone. The CT concept, in which C stands for residual ozone concentration in milligrams L^{-1} and T for contact duration in minutes, describes this efficacy on a target microorganism. As a result, the target microorganism, the environment, and the intensity of ozone therapy are all expressed in terms of CT (mg min^{-1} L^{-1}). A low ozone concentration plus an extended treatment duration is mainly equivalent to a high ozone concentration plus a shorter treatment duration for the same CT value. This equivalency hasn't been scrutinized since Haslay and Leclerc (1993) showed that high concentrations administered over brief intervals were more phytotoxic than equivalent exposures with lower concentrations applied over longer intervals. Furthermore, it was demonstrated by Finch et al. (1993) that the technique of estimating CT by using the reactant concentration after the contact time overestimates the CT required for disinfection.

The antibacterial effects of applying aqueous Ozone in two separate modes—static and dynamic—were not the same. According to Marino et al. (2018), the rate of destruction of the attached bacterial cells is, in fact, higher in dynamic settings than in static ones. This is true for all microbial species. When Ozone bubbled in water with a mixture of shredded Lettuce, high-speed stirring proved more effective than low-speed stirring (Kim et al. 1999). Furthermore, Ozone bubbling during apple washing proved to be a more successful method of sanitizing intentionally infected apples with E. coli than dipping them in ozonated water (Kim et al. 1999).

Furthermore, determining the size of the water bubbles is essential for determining how effective the disinfection process is. Reduced bubble sizes from 2.38 to 1.72 mm increased residual Ozone and microorganism inactivation for a given ozone concentration at a fixed gas flow rate (Ahmad and Farooq 1985). According to Ogden (1970), bubbles with a 0.1 cm diameter have almost 32 times more excellent contact value than those with a 1.0 cm diameter. Scientific publications have stated that water ozone is a more potent antibacterial than gaseous Ozone (Marino et al. 2018). It has been extensively documented that ozone gas inactivates bacteria only in environments with high relative humidity. According to Ishizaki et al. (1986), Ozone loses its bactericidal effectiveness at 50% or lower, and the ideal relative humidity for a gas is between 90 and 95%. Since ozone gas is highly effective at high relative humidity levels, it is advantageous for fruit and vegetable sanitation in environments where the relative humidity is often greater than

80% (Han et al. 2002). The mild effect of gaseous Ozone is only attributable to its mechanism of action, which necessitates the presence of water. Theoretically, the effectiveness of gaseous Ozone is enhanced by an increase in the relative humidity of the gas (Marino et al. 2018). While extrinsic factors significantly impact ozone efficiency, authors should not undervalue the significance of intrinsic factors.

3.3.2   Intrinsic Parameters

Table 3.1 lists the intrinsic characteristics of vegetables and their microbial population that can influence the effectiveness of Ozone for decontamination. The microorganism type, physiological state, concentration, and stress significantly impact Ozone's antimicrobial impact (Kroupitski et al. 2009; Wani et al. 2016). The age of the cells can also affect the susceptibility of the cells to ozone inactivation (Gibson et al. 2019). Wani et al. (2016) observed that for Pseudomonas spp., older colonies (7, 10, and 12 days old) were more resistant to gaseous Ozone than cells from younger colonies (2 and 4 days old). Furthermore, according to these authors, Pseudomonas sp. cultured in a refrigerator exhibit improved ozone resistance. Before P. syringae's viability declines, Ozone causes bacterial aggregation and noncultivability. Unlike those easily exposed, microorganisms embedded in surface imperfections are more protected against Ozone (Kim et al. 1999). For instance, following a fake E. Coli contamination of lettuce leaves, the intact leaf surface responded better to chlorine treatment for inactivation than the trichrome, stomata, and cut edges of damaged lettuce leaves (Seo and Frank 1999). Additionally, undamaged tissue was far more affected by the sanitizer's action than sliced tissue that had been artificially contaminated with Salmonella, as noted by Kroupitski et al. (2009). The bulk of cells were found in the cut-edge regions, preferring the injured tissue, and these cells adhered to the cuticle of the undamaged leaf surface. Wani et al. (2016) reported that the bacteria preferred adhering to the epidermal cell edge and generated sizable aggregates when employed in their study. Bacterial adhesion and exopolymer synthesis resulted in the formation of microcolonies and biofilms on leaf surfaces. According to confocal pictures of ozone-treated leaves, two to three viable cells survived in microcolonies surrounded by dead cells (Wani et al. 2016). Nonetheless, 10% of the bacterial viable counts on the leaf surface comprised individual surviving bacteria (Wani et al. 2016). As a result, according to Mah and O'Toole (2001), cells in microcolonies and biofilms on leaf surfaces may be able to withstand ozone treatment through both physical defense and the presence of improved resistance mechanisms in the biofilm bacterium. Another important factor relating to the effectiveness of ozone treatment is the product's aw. A powdered food-grade substance with variable aw was treated with gaseous Ozone at 200 ppm by Kim et al. (1999). Upon reaching 0.95 for the product's aw, over 2 log CFU/g were rendered inactive. At a comparable ozone concentration, the microbiological load of products was unaffected by an aw of less than 0.84. Ozone was equally efficient in reducing the microbial load whether the product's aw climbed from 0.84 to 0.95 since it was naturally present in a high aw product. In any tested ozone concentrations for 30 min, Sarron et al. (2021) found no discernible impact of gaseous ozone concentration on fresh and lyophilized G. stearothermophilus spores held at varying aw between 0.06 and 0.98. Furthermore, in 25 min, spore counts were lowered by 5.5 CFU/mL following an aqueous treatment (aw = 1) at 3.8 g/Nm^3 of a spore suspension. Food surface properties (nature, chemical composition, texture) and microbe characteristics (type, contamination load, and degree of adhesion) play a significant role in Ozone's inactivation of food microorganisms. For items with smooth and unbroken surfaces, like apples (Achen and Yousef 2001), tomatoes (Bermúdez-Aguirre and Barbosa-Cánovas 2013), and green peppers (Alexopoulos et al. 2013), the application of aqueous Ozone produced promising results with a low ozone demand. These products make it simple for the sanitizer to contact the bacteria directly. Microbes ought to be able to separate from plant tissue with ease. In contrast to the stem-calyx region, which is irregular and has spaces for bacteria to hide, the surface of an apple is smooth, uniform, and easily exposed to Ozone (Achen and Yousef 2001). Contrary to Kim et al. (1999), however, a study by Alexopoulos et al. (2013) showed that Lettuce has a very uneven and rough surface with lots of hiding areas that could serve as a bacterial niche. Microbial inactivation appears more difficult when the surface is more complex regarding porosity and roughness, such as on the roots of carrots (Bermudez-Aguirre and Barbosa-Canovas 2013). Carrots' porous surface, which shields bacteria from ozone treatment, significantly impacts the inactivation of germs. As such, it is critical to guarantee that the target bacteria and Ozone come into direct contact. A range of application techniques are used in this context to improve the quality of treated items, such as washing, dipping, stirring, and bubbling. Microorganisms are destroyed by Ozone by the gradual oxidation of essential biological components. The main target of Ozone is the surface of bacteria, which can cause cell wall ruptures and subsequent cellular disintegration due to oxidation. Since their outer membrane is composed of lipoproteins and polysaccharides, Gram-negative bacteria, like E. coli, have thin peptidoglycan lamella, making them more vulnerable to Ozone at the beginning of ozone therapy. In comparison to Gram-negative bacteria, Gram-negative bacteria have greater D-values. According to Evrendilek and Ozdemir (2019), bacterial destruction was strain-related rather than Gram-related when the ozone treatment was extended.

3.4   Effect of Ozone Treatment on the Quality of Essential Vegetables

We will now discuss the function and effects of gaseous and aqueous ozone treatments on three distinct food matrices: fresh carrots, Lettuce and salads, and tomatoes while debating the usage of Ozone in the vegetable business.

3.4.1   Effects of Ozone Treatment on Carrot Quality

According to Yahia (2019), carrots rank among the top ten vegetable crops farmed globally in terms of economic importance. Almost 37 million tonnes of carrots were grown for human consumption worldwide 2012. Asia was the region that produced the majority of the world's carrots (61.8%), followed by Europe (22.6%) and America (9.1%). Carrots with an orange hue are famous all around the world. The phloem, the pulpy outer cortex, and the xylem, the inner core, comprise most of the taproot. Superior carrots have a higher percentage of cortex than the core. Vitamins, dietary fiber, and phytochemicals called carotenoids are many physiologically active compounds among carrots. Fresh carrots are becoming more and more popular since they are known to be a significant source of naturally occurring hydrophilic and lipophilic antioxidants, including lutein, lycopene, and chlorogenic acids, all of which have anticancer properties (Sharma et al. 2012). It is beneficial to have a high initial total viable count of 5–7 log CFU g^{-1} in low-acid settings (pH 6.0–6.5) to facilitate a quick expansion in the microbial population. Because of their limited shelf life, fresh carrots should be eaten within a few days, which will reduce their market potential and raise issues with microbiological safety. Nonetheless, fresh carrots can be stored for up to five months at 0 °C and high relative humidity (98–100%) (García-Gimeno and Zurera-Cosano 1997). However, excessive decay brought on by bacteria such as Sclerotinia sclerotiorum and Botrytis cinerea, as well as the emergence of bitterness and the loss of texture and flavor, can all contribute to a deterioration in the quality of carrots. In this case, Ozone is the technology that can, first, prolong the shelf life of fresh carrots found in stores after a step to lessen microorganisms during washing and, second, lessen decay during a lengthy storage period while maintaining the quality of stored carrots.

3.4.2   Effect of Continuous Gaseous Ozone Exposure on the Quality of Stored Carrots

A continuous ozone treatment aims to extend fresh whole carrots' shelf life and storage period immediately following harvest. We will start by concentrating on the carrots' appearance, which is, as previously stated, the primary factor in consumer purchases. In every study, the impact of Ozone on quality metrics like color was discussed. The product's appearance is crucial since any color change could be interpreted as a sign of aging. Following treatments at 450 ppb (0.45 mg L^{-1}) for 48 h [69], 7.6 mg L^{-1} for 15 min (Singh et al. 2002), and between 1 and 5 mg L^{-1} for 9.5 to 110.5 min, no discernible change in carrot color was seen. However, the ozone treatment resulted in some damage that showed up as bleaching (Bridges et al. 2018), scattered blotches of slightly discolored, brown periderm at 50 nL L^{-1} for six months (Hildebrand et al. 2008), and dry white blotches at 60 μL L^{-1}, eight h daily, for 28 days (Lewis et al. 1996). These harmful out-comes suggest that the oxidative stress caused by Ozone causes severe physiological harm to carrots. Carrots become discolored white when their surface dehydrates (Amanatidou et al. 2000). Color changes and surface pitting in ozone-treated carrots may also impair consumer appeal. Most published research found that the ozone treatment did not affect carrots' physical and physiological quality immediately after or during the storage period for the evaluated concentration and treatment period. We will individually review each of the writers' claims regarding the detrimental effects of Ozone on fresh carrots. One significant rheological characteristic relevant to fresh carrots is firmness, which is linked to weight loss. Firm-textured carrots are an indicator of their freshness and healthfulness. Numerous investigations revealed that Ozone has no impact whatsoever on the hardness of carrots. Firm-textured carrots are an indicator of their freshness and healthfulness. Numer-ous investigations revealed that Ozone does not affect the toughness of carrots whatsoever (Luwe et al. 1993; Marino et al. 2018). Conversely, Forney et al. (2007) reported a delay in tissue toughening and a decreased stiffness due to ozone treatment. Because of the reduced lignification of cell walls was linked to changes in the concentration of cellulose, lignin, and hemicellulose (Chauhan et al. 2011). Furthermore, the high oxidation power of gaseous Ozone during postharvest treatment and storage encourages other undesired changes in carrot quality. Increased respiration rates, electrolyte leakage, and sucrose con-tent were signs of physiological disturbances (Forney et al. 2007). According to Forney et al. (2007), the presence of terpenes and hexanal in the headspace suggests that lipid oxidation has occurred and that this treatment might improve the flavor of carrots. Accord-ing to Forney et al. (2007), Ozone is a postharvest stressor that increases respiration and ethanol synthesis. The aberrant metabolism brought on by a rise in ozone concentration is the cause of these increased respiration rates (Liew and Prange, 1994a, b). With the most minor physical and physiological harm possible, a 28-day delivery of 15 μL L^{-1} of Ozone offers some protection against disease (Liew and Prange 1994a, b).

Regarding the microbiological quality of carrots, most research found that longer expo-sure times and higher ozone concentrations are associated with increased bactericidal action (Bridges et al. 2018). At higher concentrations, ozone treatment does, however, degrade the color qualities of carrots at harvest. Since Ozone mainly affects the outer surface of the roots, it would hinder bacteria, which are primarily found on the surface of the core. In addition to changing the surface physically, this surface treatment may cause minor changes in the carrots' pH, soluble solids (SS), glucose, fructose, sucrose, and galactose. Furthermore, carrots have a longer shelf life thanks to Ozone. Ozone was fungistatic rather than fungicidal on B. cinerea and S. sclerotiorum (Hildebrand et al. 2008). According to Hildebrand et al. (2008), the treatment included raising the concen-tration of isocoumarin, which helped to slow down the rate at which these two altering microorganisms expanded the lesion.

Furthermore, E. Coli O157:H7 (Singh et al. 2002), Shiga toxin-producing E. Coli (STEC), Salmonella enterica, and Listeria monocytogenes (Bridges et al. 2018) were all

susceptible to the bactericidal effects of gaseous Ozone, which grew stronger with exposure duration and concentration. We have discussed how Ozone might improve carrots'
nutritional value, visual appeal, and sensory appeal.

We now need to concentrate on the carrot processing conditions. All the writers mentioned that the ozone concentration in the gas was measured at the start of the treatment
when it entered the reactor containing the carrots. Applying 60 μL L^{-1} of Ozone produced
a distinct residual ozone concentration at 2 and 16 °C, according to Liew and Prange
(1994a). They also found that temperature and applied ozone concentration impact residual ozone concentration. In fact, at 2 and 8 °C, they detected a more extensive residual
ozone content than at 16 °C.

Furthermore, a notable distinction was seen between the concentrations used in the
application and the residual concentration. To determine the amount of Ozone lost due to
various phenomena, such as dilution in the treatment container's volume, reaction with
plant compounds like pesticide residues or microorganisms, duration of the purge, etc., it
is imperative to report the ozone concentration in the reactor at the start, during, and end
of the treatment. It's crucial to remember that the writers' units are not all the same.

Furthermore, μL L^{-1} was the unit of measurement in the work of Liew and Prange
(1994a). An ambient ozone analyzer model IN-2000–5 was used in this work to measure
the ozone concentration using UV absorption. This model is intended to provide ozone
concentrations in parts per million (ppm) between 0 and 1000 ppm. Forney et al. (2007)
and Hildebrand et al. (2008) treated carrots with the same Simpson Environmental Cor-
poration generator using a comparable unit (nL L^{-1}). However, according to Sharpe et al.
(2009), this generator produced an ozone concentration of parts per billion when carrots
were treated. The terms "parts per billion" (10^{-9}) and "parts per million" (10^{-6}) denote
the units "ppb" and "ppm," which indicate ratios. Whether these units correspond to volume, mass, molar concentration, or massic concentration is not specified. Authors should
use international units.

Furthermore, properly harmonizing the units is essential for the reader to replicate
the entire treatment under identical conditions and effectively compare the various treatments. To create comparable working units, Bridges et al. (2018) reported the features
of the ozone concentration in mg/m3 and a processing rate represented in μg O3 g^{-1} of
produce. The supplementary data suggested that the 0.86 and 1.71 μg O$_3$ g^{-1} treatment
dosages facilitated the comparison across multiple scholarly publications. Unfortunately,
this processing rate cannot be determined in all scientific studies because of the severe
lack of information in scientific journals (e.g., no stated flow rate, absence of exact quantity of treated product, etc.). As demonstrated previously, there is a great deal of variation
in the concentrations of Ozone, their application times, methods, and environmental factors like pH, temperature, and humidity. All of these conditions are pretty effective and
involved a significant increase in microbial quality as well as relative conservation of
the physical and visual attributes of the stored carrots, despite the high variability of the
treatments (e.g., in terms of CT, it comprised between 1.73 and 804.6 mg min^{-1} L^{-1}).

However, more research is required to provide the ideal ozone concentration under perfect circumstances (time, temperature, RH, flow rate, etc.) to prevent deterioration and preserve quality with the least physiological and physical harm—foods 2021, 10, 605 13 of 39.

3.4.3 Lettuce and Salads

Ready-to-eat veggies have become increasingly popular over the last 20 years, primarily due to their convenience and health advantages, and this trend shows no signs of abating. Over 80% of the fresh-cut produce comprises salad bars, supermarkets, and fast food restaurants. Fresh-cut salads are one of the commodities in higher demand due to their importance to human diets. According to Baslam et al. (2013), Lettuce is a good source of minerals, including calcium and iron, vitamins A and C, and phytochemicals with significant antioxidant potential, such as phenolic antioxidants. Mesophilic bacteria are found in relatively low concentrations in the inner leaves of Lettuce, typically 104 CFU g^{-1}. However, the quantities of these bacteria in the packaged product are significantly higher because of contamination from unit operations applied from the farm to the fork pathway (preparation, handling, cleaning, trimming, washing, drying, packaging, storage, and transport), especially the shredder (Garg et al. 1990). Since lettuce leaves are typically eaten raw, postharvest handling is crucial to preventing foodborne pathogen contamination. Disinfection is one of the most critical processing stages affecting salads' safety, quality, and shelf life.

Furthermore, eating fresh produce that contains leafy greens may increase the risk of spreading known foodborne infections, as these veggies are linked to 22% of illnesses. Salmonella, Listeria, Escherichia coli, Bacillus cereus, Campylobacter jejuni, Staphylococcus aureus, and Shigella are pathogenic bacteria that frequently contaminate Lettuce. These bacteria can attach to open stomata, fissures in the cuticle or trichome, and the leaf epidermal cell margin (Hassenberg et al. 2007; Karaca and Velioglu 2014). According to a WHO study, leafy green vegetables are the commodity group that should be of the utmost concern regarding foodborne outbreaks. Furthermore, a maximum lactic acid bacterium count of 106 CFU g^{-1} indicates the start of ready-to-eat salad spoiling, according to Garcia-Gimeno and Zurera-Cosano (1997). Their estimates for a product's shelf life suggested that it may last up to 8.7 days in storage at four °C, which is longer than the producers' recommended 6-day period. This shelf life exceeds the current safety guidelines for fresh-cut Lettuce, which should last between five and seven days. In the fresh-cut sector, chlorine is typically used to keep Lettuce from being contaminated and to prolong its shelf life. However, ozone treatment is a sustainable technology that can enhance the overall quality of Lettuce and salads.

3.4.4   Effect of Continuous Ozone Exposure on Quality of Salads Stored Lettuce

Ozone in the gaseous phase is the first system used to apply Ozone to improve the quality of stored Lettuce because it is known that ozone molecules have a longer half-life in air than in aqueous solution and a higher diffusion rate. This application has been studied under a variety of conditions, including hydroponically grown butterhead lettuce seeds and seedling crops (Kleiber et al. 2017), in greenhouse growing of four-week-old Lettuce (Calatayud and Barreno 2004), during a storage period right after harvesting (Galgano et al. 2015), and for ready-to-eat leafy vegetables (Wani et al. 2016). Butterhead lettuce seeds treated with Ozone at a rate of 14 g h^{-1} for 30 min each day exhibit enhanced germination, increased uptake of elements, improved chemical composition and chlorophyll content, improved physiological processes, increased growth, and ultimately, increased yield. According to Kleiber et al. (2017), a second ozone treatment of seedlings degrades plant health and reduces production under outdoor circumstances in a manner akin to that of tropospheric Ozone. In addition, Calatayud and Barreno (2004) noted that in two of the lettuce kinds they tested, ozone treatment (8.2–83 nL L^{-1}, 12 h day^{-1}, 60 days) changed growth, reduced mean weight and subsequently productivity, and decreased its market value. They found notable distinctions between the Valladolid and Morella kinds, with the latter being more vulnerable to ozone depletion because they lack the antioxidant anthocyanins (Calatayud and Barreno 2004)... In general, ozone gas treatment of food goods can be accomplished by introducing small amounts of Ozone (0.1–10 µL L^{-1}) into the storage atmosphere. When Ozone is utilized as a gas immediately after harvest, it is exposed for longer lengths (ranging from days to months) than when dissolved in water. Galgano et al.'s research from 2015 demonstrated that applying Ozone for seven days at four °C at low quantities (0.2 ppm) did not affect the sensory attributes of iceberg lettuce, such as color stability.

Additionally, total mesophilic bacteria and total coliforms were inhibited by Ozone, whereas yeasts, molds, E. coli, and pseudomonas were very occasionally found (Galgano et al. 2015). Compared to only refrigeration, these authors found that the application of Ozone is efficient in controlling microbial growth during raw material chilling storage durations. Applying 1 µL L^{-1} on iceberg lettuce for 10 min demonstrated a reduction in target microorganisms and did not change color after processing similarly (Wani et al. 2016). Additional therapies comprised of high ozone dosages (ranging from 100 to 10,000 µL L^{-1}) were administered briefly (Karaca and Velioglu 2014). These treatments aimed to extend the shelf life of lettuces at 4 °C. Both microorganisms were reduced by 1.0–1.5 logarithmic units, but significant losses were observed in critical bioactive compounds, such as antioxidant activity, total phenolic contents, and ascorbic acid. Regardless, it's crucial to note that the food business seldom uses these gaseous ozone treatments. Washing with aqueous Ozone is preferred since green salads and Lettuce have a limited shelf life.

3.4.5   Tomatoes

A highly favored crop, tomatoes are grown in over 100 countries, with 244 million tonnes produced globally in 2018. According to Cantwell (2000), the three most crucial quality standards for tomatoes are their red color, a firm yet juicy texture, and a nice flavor. Due to their high nutritional content, including lycopene and ascorbic acid, and their anti-inflammatory, antioxidant, and anticancer properties, tomatoes are ingested (Salehi et al. 2019). On the other hand, tomatoes are susceptible to deterioration by microorganisms, especially fungus, during storage and can become infected with foodborne diseases like Salmonella or Norovirus. Tomatoes are often picked by hand and placed into boxes. They are then delivered to packaging houses, where the fresh fruit is slightly treated before storage. Physical treatments that would eradicate Salmonella and Norovirus are not applied to tomatoes. Good manufacturing and farming procedures are necessary to prevent these pollutants. The quality of the wash water and the storage environment are the two primary elements that regulate the growth of microorganisms. You can apply Ozone while storing or cleaning.

3.4.6   Effect of Exposure to Continuous Gaseous Ozone on the Quality of Stored Tomatoes

For tomatoes to ripen and have a longer shelf life, storage is a crucial step. Because of their high water and nutrient content, Tomatoes are highly susceptible to microbial deterioration, especially from yeasts and molds. Furthermore, several abiotic stressors encountered during the ripening and harvesting phases are linked to the shelf life of tomatoes (Alenazi et al. 2020). Tomatoes are often preserved by storage in a positive pressure chamber; however, this affects the fruits' functional and organoleptic qualities and the hardness of their flesh (Alenazi et al. 2020). Moreover, water that can promote microbial proliferation is released due to gas exchanges (such as CO_2, O_2, and ethylene) during ripening and storage. According to Smith et al. (2007), this water can cause a variety of dangerous bacteria and fungi to internalize. This is especially true for stem scar tissue as opposed to smooth tissue. Gaseous ozone treatment can be employed before or during storage, especially when refrigeration is not an option, to prevent these developments. Numerous authors have investigated the use of Ozone to decontaminate tomatoes (from viruses or fungi) while preserving the nutritional value and organoleptic characteristics of various varieties of mature red tomatoes and ripened green tomatoes or packed tomatevera types of ripe tomatoes, including cherry, beefsteak, and grape tomatoes, have been tried. Salmon lla enteritidis was injected onto the surface of cherry tomatoes at low and high doses (3 and 7 log CFU tomato^{-1}) (Das et al. 2006). Ket for a day or two to promote bacterial adherence. In cases where tomatoes were treated with 10 mg L^{-1} for 5 or 10 min, no bacteria were found after 1 or 4 h of storage for the low inoculum. Complete bacterial killing was attained for the high inoculum after 15 min at 20 mg L^{-1} for a 4 h attachment time. A high r dosage of 30 mg L^{-1} was utilized to shorten the duration, but this resulted in a shift from red to yellow. In a study conducted by Bridges et al. (2018), the efficiency of gaseous Ozone in eliminating a combination of pathogens (S. enterica, Escherichia coli (STEC), and Listeria monocytogenes) was examined when the final contamination level on beefsteak tomatoes exceeded 6.5 log CFU g^{-1}. Following a 5-h exposure period at 1.71 μg O_3 g^{-1} product, E. coli STEC showed a maximum reduction of 1.6 log CFU g^{-1}, while Salmonella and Listeria showed a 1.1 log CFU g^{-1} decrease. Though the extended treatment period and higher concentration allowed for more bacterial eradication, regrettably, a bleaching of the tomato epidermidis was seen. Differ nt amounts of Ozone (1.71, 3.43, and 6.85 mg L^{-1}) were applied to grape tomatoes for two or four hours, depending on whether they had been inoculated with Salmonella on their smooth surface and scar stem or not (Wang et al. 2019). A 2-lo CFU $fruit^{-1}$ reduction in Salmonella was seen at 6.85 mg L^{-1} for 2 h, regardless of the tissue type (smooth zone or stem scar). In lig t of the native microbiota, routinely assessed during storage at 10 °C, a decrease in the overall plate count was noted on days 1 and 7 of storage for concentrations of 3.43 and 6.85 mg L^{-1}, respectively. However, the yeast and mold populations remained unaffected despite the gaseous ozone treatment. Additionally, during these tests, off-note aromas and visual deterioration were observed. Following a 4 h exposure to a dosage of 6.85 mg L^{-1}, the tomatoes showed signs of moisture, indicating a rupture in the skin. During storage, two nutritional markers were monitored. Day 1 saw no change in the levels of lycopene or ascorbic acid. Nevertheless, a steady reduction in these molecules was seen after storage, and by day 21, only one-third of the ascorbic acid was still present. Degradation of lycopene was associated with changes in red hue. In an effort to enhance their findings, the authors conducted additional tests with 800 and 1600 ppm for 30 min following Salmonella inoculation on the smooth surface and scar stem, either in conjunction with hydrogen peroxide or not (Fan et al. 2020). When one gas was used alone, only a 0.5 log CFU fruit1 reduction was seen; however, when aerosolized hydrogen peroxide was added to the ozone gas treatment, a 5.2 log CFU fruit1 reduction was seen on the smooth surface.

In their 2013 study, Tzortzakis et al. concentrated on the effects of Ozone on fungal decomposition as well as the nutritional value and sensory attributes of fully ripe tomatoes inoculated with Botrytis cinerea and exposed to either low-level ozone enrichment (0.1 μmol mol^{-1}) at 13 °C or charcoal-filtered clean air. The de-elopement of apparent lesions in all treated fruit and a significant reduction in B. cinerea spore production and viability were the outcomes of ozone enrichment. Regarding tomato quality, there was no effect on weight loss, antioxidant status, CO_2/H_2O exchange, or the content of organic acids, total phenol, or vitamin C under any of the tested conditions (0.005 to 1.0 μmol mol^{-1} Ozone at 13 °C and 95% relative humidity) (Tzortzakis et al. 2007a). According to sensory research, tomatoes treated with 0.15 μmol mol^{-1} Ozone were more appreciated than those treated under other circumstances. The authors proposed that Ozone may be used instead of insecticides to prevent fungal growth during storage.

Occasionally, mature-green tomatoes are harvested and allowed to ripen at room temperature and humidity without management. These circumstances encourage the growth of fungi; thus, utilizing an ozone disinfectant could assist in maintaining quality. Zambre et al. (2010) used two indices to assess ripening under ozone treatment at varying temperatures: the rotting index, which was based on 75% of maximal spoiling, and red color development. After being placed in an ozone chamber and treated for 10 min at 20, 35, and 50 ppm, tomatoes at various stages of ripening were individually packed in open bags and stored at 15, 20, and 35 °C with $68 \pm 3\%$ relative humidity. The most significant outcome was a 10-day shelf life extension with a 3.6 day ripening delay when the tomatoes were treated at 35 ppm for 10 min and stored at 15 °C.

Nevertheless, the advantages of ozone therapy were utterly destroyed by a rise in storage temperature. A long r shelf life was produced by ozone treatment and lower temperature, which lowered the initial microbial count and slowed the rate of microbial development. In the 2010 study, Venta et al. assessed the effects of gaseous Ozone on a few physical–chemical characteristics and loss in unripe tomatoes in Cuba after harvest. Green tomatoes with lower lycopene and ascorbic acid concentrations than the control showed the most outstanding results regarding firmness, weight loss, and spoiling when exposed to 25 mg m^{-3} Ozone for 2 h per day for 16 days. Tomatoes treated with Ozone had a longer shelf life (only 14% damaged fruit compared to 54% for the control group), most likely because it slowed down the rate of ripening.

On the other hand, the harm associated with ozone treatment is noted when a high dose is used. Finding the ideal circumstances is necessary to raise tomato quality. It has already been reported (Tzortzakis et al. 2007b) that Ozone alters the ripening process, and proteomics analysis can explain the changes in physiological state. The pr tein profiles of tomatoes were compared by Tzortzakis et al. (2007b) after they were stored for one week in each of the following four conditions: Ozone (0.05 μmol L^{-1}), wound inoculation with B. cinerea (a 2.5 mm mycelial plug in a wound), and with or without treatment (charcoal-filtered clean air) following two pretreatments. These pretreatments aimed to assess any possible memory effect associated with the ozone treatment. When tomatoes were removed from the ozone-enriched environment, the increase in protein content caused by ozone treatment was reversed. When t tomatoes were exposed to Ozone and B. cinerea wound inoculation, their proteomes were altered; however, when tomatoes exposed to Ozone were placed in clean air, or when tomatoes inoculated with B. cinerea were placed in an ozone atmosphere, their proteome shifts were qualitatively repressed. This finding was clarified because oxidative stress proteins are produced by Ozone, which primes tomatoes to react to infections.

Additionally, specific proteins linked to ethylene synthesis are downregulated when exposed to Ozone. As a result, a reduced ethylene rate would lessen ripening and slow the growth of infections. An in-package ozone treatment device can treat tomatoes (Fan et al.

2012). Salmon lla Typhimurium, E. Coli O157:H7, or L. innocua-inoculated tomatoes were sealed in a bag and placed within a treatment chamber to produce Ozone inside the bag. A decrease ranging from 1.8 to 6 log CFU unit^{-1} was noted, contingent on the strain and the region under consideration (surface or scar stem). The pa Kage's ozone treatment phase did not significantly impact the tomatoes' firmness or color after 22 days at 22 °C. Several studies have shown that Ozone can be used to clean and store tomatoes. This has the advantages of increasing the shelf life and lowering the number of microorganisms. Suppose we wish to preserve the product's sensory and nutritional features. In that case, we must consider the stages of development, environmental factors like temperature and humidity, and the type of tomatoes used (weight of the product). Ripening can be controlled with ozone gas. At the same time, users may find this highly intriguing since it allows them to postpone the tomato ripening process and sell tomatoes just when needed.

3.5   Effect of Ozone Treatment on the Quality of Essential Fruits

Ozone's ability to impede postharvest metabolic activities is how it prolongs fruit quality. One of the main metabolic processes that cause quality degradation is ethylene synthesis (Huyskens-Keil et al. 2012)—specifically, knowing how ethylene biosynthesis works and the variables that influence it can explain it. Methio ine dissociates under normal circumstances to create S-Adenosyl Methionine (S-AdoMet.). S-aden sylmethionine (SAM) synthetase catalyzes this process, which is driven by adenosine triphosphate (ATP) (Wang et al. 2002). One AT molecule is used to make one S-AdoMet molecule (Wang et al. 2002; Jouyban 2012). Nuclei acids, proteins, and lipids are just a few biological components for which the S-AdoMet serves as a methyl group donor. Although h A. C. C. is a direct precursor of ethylene, it is often referred to as a precursor of the polyamine production route (Jouyban 2012). Wang e al. (2002) state that the conversion of S-AdoMet to ACC, which ACC Synthase catalyzes, is the rate-determining step in the biosynthesis of ethylene. In order to maintain methionine in the cell, methylthioadenosine (MTA), a byproduct of ACC production, is recycled back into methionine. By removing the ACC pool, the malonylation of ACC generates many-ACC (MACC), which lowers ethylene synthesis (Ijaz 2016). The enzyme known as ACC oxidase catalyzes the last stage of ethylene biosynthesis, in which ACC is utilized as a substrate to produce cyanide, carbon dioxide, and ethylene (Jouyban 2012; Ijaz 2016).

3.5.1 Ozone Inhibits the Activities of A.C.C. Synthase and A.C.C. Oxidase, and Directly Oxidizes Ethylene

According to Minas et al. (2014), Ozone inhibits the production of ACC synthase, which stops ACC from building up and inhibits ethylene biosynthesis. Wang et al. (2002) referred to this crucial stage in the ethylene synthesis process as a rate-determining step. According to a study by Minas et al. (2014), ozone treatment suppressed the activity of the enzyme ACC synthase throughout the storage period, which resulted in lower ACC content in ozone-treated kiwifruit cv. "Haywa d" compared to the untreated, delaying the onset of ripening. This indicated that this process might significantly influence ethylene production since it controlled A.C.C., an ethylene precursor.

Additionally, Ozone decreased the enzyme ACC oxidase's activity, which catalyzes ACC's oxidation to produce carbon dioxide, cyanide, and ethylene. Ozone's ability to suppress the expression of the receptor gene AdACO1, which codes for the activity of the ACC oxidase enzyme, may be related to its power to suppress ACC oxidase enzyme activity. When the enzyme's activity was inhibited, less ethylene was produced, which led to a lower respiration rate than in fruit that wasn't treated (Toti et al. 2018)—melons cv. "Calde " softened due to the substantial reduction in ethylene synthesis caused by using Ozone to decrease ACC oxidase activity. The phrase that determines the rate of ethylene synthesis is the inhibition of ACC synthase and ACC oxidase activities (Yin et al. 2013). To improve the quality of fresh food, Ozone must block the activity of enzymes that aid in ethylene biosynthesis, suppressing ethylene formation. This is because an increase in ethylene synthesis triggers several metabolic reactions, including respiration, which results in softening, a reduction in nutritional value, and a general decrease in quality. According to Smilanick (2003), Ozone is very reactive and efficient at eliminating ethylene gas from packhouses, packages, and storage areas. It directly produces carbon dioxide and water when combined with ethylene gas in the storage room atmosphere (Yin et al. 2013). According to Skog and Chu (2001), the presence of gaseous Ozone causes ethylene to oxidize, producing carbon dioxide and water. Ozone is a potent oxidant; when it is in its aqueous state, a single oxygen molecule that the ozone releases attaches itself to other water molecules, oxidizing them into new compounds. All organic and many inorganic compounds will react with a single oxygen molecule and become oxidized. The organic chemicals continue to separate from the water after oxidizing. The ability of Ozone to degrade and oxidize anything it touches is known as oxidizing power. Thanks to this property, it has become one of the most potent disinfectants and oxidizers. When O one and ethylene combine to form CO_2 and water, the amount of ethylene in the air surrounding fresh produce is reduced, slowing the deterioration of fresh produce quality and slowing down metabolic processes.

3.5.2   Ozone Inhibits the Activities of Cell Wall Degrading Enzymes

Ozone treatment has demonstrated that fruit softening is much reduced; however, the alterations in cell walls accompanying this process are not fully understood (Rodoni et al. 2009). According to Toti et al. (2018), ethylene biosynthesis is the primary process linked to a decline in fruit quality and incredibly softening. The authors went on to explain that fruit softening is related to the activity of enzymes that break down cell walls, including pectin methylesterase (PME) and polygalacturonase (PG)—fruit softening results from the activation of these enzymes by metabolic processes like ethylene generation. Toti et al. (2018) went on to explain that melons exposed to Ozone had a drop in the activity of three enzymes that break down the cell wall: polygalacturonase, α-arabinopyranosidase, and β-galactopyranosidase. They did not, however, find any evidence of a significant impact of Ozone on pectin methylesterase activity.

Additionally, Minas et al. (2014) found that PG activity controls the critical influence of Ozone in regulating kiwifruit softening. These results suggest that cell wall deterioration and fruit and vegetable softening are related. These results indicate that ozone treatment delays fruit softening by blocking the actions of enzymes that break down cell walls, primarily pectin methylesterase (PME) and polygalacturonase (PG). This preserves the firmness of fresh food. Since ethylene regulates the biochemical processes involved in fruit ripening, especially on climacteric fruit, adequate ethylene production may be necessary for ripening, and it explains why Ozoneozone can block the activities of cell wall disintegrating enzymes. Through the suppression of gene expression in cell wall degrading enzymes, such as AdACS1 and AdACO1, Ozone indirectly inhibits the activity of these enzymes, which are required for the manufacture of ethylene. Ripeni g and softening are delayed when ethylene production is decreased due to suppression of ACS and ACO activities. Enzymes that break down cell walls, such as PG and PME, are directly inhibited by metabolic processes like ethylene synthesis. By decreasing the rate of pectin solubilization, Ozone directly inhibits the activities of enzymes that break down cell walls, which in turn suppresses the activities of galactosidase (β-Gal) and polygalacturonase (PG). As a result, the rate of depolymerization and polymethyl esterase (PME) activity further decreased, which decreased the rate at which the fruit treated with Ozone disassembled its cell wall.

3.6   Factor Affecting the Shelf Life of Ozone Treatment of Fruits

The two central pillars of a sustainable food system are lowering postharvest losses and preserving the quality of fresh produce (Opara 2013). The high rate of food waste and postharvest losses has led to a significant global issue with food insecurity. A cons mer does not receive about 20–50% of fresh product harvested due to postharvest losses between harvesting and eating (Kader 2004; Hodges et al. 2011). Ozone, a popular treatment that inhibits the biochemical processes linked to the deterioration of fruit quality without leaving any residues on the fruit surface, can be used to minimize postharvest losses (De Alencar et al. 2013; Tran et al. 2013). According to studies, Ozone's mode of action revolves around its capacity to maintain antioxidants, which are essential for fruits and vegetables as a defense mechanism. By reacting with ROS, antioxidants shield fresh horticulture produce and play a significant role in the human diet following consumption of the fresh product (Bill et al. 2014).

Ozone has been shown in numerous studies to be beneficial in preserving fruit and vegetable quality and prolonging their shelf life. For example, postharvest treatment of strawberry fruit (Fragaria ananassa Duch. cv. Camarosa) and storage at two °C in an ozone-containing storage environment (0.35 mg/L) led to shelf life extension and quality maintenance. According to Perez et al. (1999), this was explained by Ozone's ability to preserve sugars and up to three times the ascorbic content of untreated fruit. According to Pak and Dixon (2001), immersing 'Hass' avocado fruit in ozonized water significantly decreased ripening rots, which appear as stem end and body rot. This is explained by the ability of aqueous Ozoneozone to suppress microorganisms linked to microbial development, which accelerates stem end and body rot. The fruit softening, weight loss, and decay incidence of strawberry fruit treated with Ozone ($1.5 \mu L \, L^{-1}$) and held at two °C for three days before being moved to room temperature were significantly reduced, according to Nadas et al. (2003). According to Whangchai et al. (2006), the microbial population on the fruit surfaces significantly decreased after 60 and 120 min of exposure to 200 μL L^{-1} Ozone on longan fruit.

Pears were exposed to gaseous Ozone (100 mg/L) for 60 min at a flow rate of 2.3 L min^{-1}, and then they were stored at 25 °C, which extended their shelf life and inhibited microbiological development. This was explained by Ozone's ability to dramatically lower the bacteria counts in treated fruit compared to the untreated control group (Alencar et al., 2014). According to Minas et al. (2014), applying ozone ($0.5 \mu L \, L^{-1}$) to kiwifruit postharvest resulted in a notable delay in fruit ripening, preservation of quality, and an extension of shelf life at 20 °C. This was ascribed to Ozone's ability to suppress the primary enzymes that catalyze the synthesis of the precursors of ethylene, 1-aminocyclopropene-1-carboxylate synthase (ACC synthase, ACS), and Aminocyclopropanecarboxylate (ACC) oxidase. According to Minas et al. (2014), Ozone inhibits the activity of ACC synthase (ACS) and ACC oxidase (ACO), two essential enzymes that code for ethylene production. This shows that Ozoneozone has a lower influence on ethylene biosynthesis. It accomplishes this by preventing the expression of two genes, AdACS1 and AdACO1, which code for the activity of the ACS and ACO enzymes. As a result, less ethylene is produced, and the activity of both enzymes accumulates more slowly. When fruit ripened, the amount of ACS was reduced because ozone treatment prevented the ACS enzyme from building up. In con rast, a control (untreated) fruit showed a climacteric elevation in the ACS, indicating a good response. Ozone's ability to suppress and maintain low levels of ACS activity raises the possibility that ACS function is a significant regulatory mechanism for Ozone's control over kiwifruit ripening.

Additionally, ozone treatment lowered and maintained the ACO accumulation rate much lower than in the fruit that was left untreated (control). According to the results, ozone treatment may have directly or indirectly reacted with these enzymes, lowering their concentrations and influencing how they responded to fruit ripening. According to Skog and Chu's (2001) study, tomatoes stored in gaseous Ozone had a decrease in the formation of 1-aminocyclopropane-1-carboxylic acid (ACC). According to Huyskens-Keil et al. (2012), Ozone's ability to marginally lower respiration rate was linked to the preservation of asparagus's quality and shelf life. Ozone's ability to lower the rate of ethylene biosynthesis by blocking the actions of enzymes involved in the creation of ethylene precursors is the reason for its effectiveness as a postharvest treatment, according to all of the research mentioned above. According to Ansah et al. (2018), ethylene has a detrimental impact on the postharvest performance of horticulture items by speeding up the metabolic processes that cause fresh produce to deteriorate. According to Yaseen et al. (2015), after two months (60 days) of storage, apples (Royal Gala, Golden Delicious, and Fuji) stored at 1 ± 1 °C with gaseous Ozone (0.5 μL L^{-1}) had better postharvest quality. This is explained by Ozone'sozone's ability to lower the number of fungi and generate patulin. According to Palou et al. (2001), orange and lemon exposure to gaseous Ozone at 1.0 ± 0.05 mg/L at ten °C considerably preserved fruit quality and increased shelf life compared to the control treatment. After week, ozone treatment decreased the pace of infection and postponed the formation of blue and green mold. According to Alothman et al. (2010), fruit's antioxidant capacity was increased after 20 min of exposure to Ozone (8 ± 0.2 mL/s) for pineapple and banana. This study demonstrated that fresh-cut fruit's antioxidant status can be improved by using Ozone in the minimally processed fresh food business. Sachad n-Krol et al. (2016) found that pepper fruit treated with Ozone maintained higher levels of flavonoids in the pericarp tissue, estimated to be 25% more than those of the untreated control. Nevertheless, most studies' findings indicated that Ozone primarily acts by reducing the populations of microorganisms and fungi (Bermúdez-Aguirre and Barbosa-Cánovas 2013).

References

Achen, M.; Yousef, A.E. Efficacy of Ozone against Escherichia coli O157:H7 on Apples. J. Foo Sci. 2001, 66, 1380–1384.

Ahmad, M.; Farooq, S. Influence of Bubble Sizes on Ozone Solubility Utilization and Disinfection. Water ci. Technol. 1985, 17, 1081–1090.

Alenazi, M.M.; Shafiq, M.; Alsadon, A.A.; Alhelal, I.M.; Alhamdan, A.M.; Solieman, T.H.I.; Ibrahim, A.A.; Shady, M.R.; Al-Selwey, W.A. Improved functional and nutritional properties of tomato fruit during cold storage. Saudi . Biol. Sci. 2020, 27, 1467–1474.

Alencar, E.R., Faroni, L.R., Pinto, M.S., Da Costa, A.R., Carvalho, A.F., 2014. Effectiveness of Ozone on postharvest conservation of pear (Pyrus communis L.). Int. J Food Processing Techno. 5.

Ale opoulos, A.; Plessas, S.; Ceciu, S.; Lazar, V.; Mantzourani, I.; Voidarou, C.; Stavropoulou, E.; Bezirtzoglou, E. Evaluation of ozone efficacy on the reduction of microbial population of fresh cut lettuce (Lactuca sativa) and green bell pepper (Capsicum annuum). Food C ntrol 2013, 30, 491–496.

Alothman, M., Kaur, B., Fazilah, A., Bhat, R., Karim, A.A., 2010. Ozone- nduced changes of antioxidant capacity of fresh-cut tropical fruits. Innov. Food Sci. Emerg 11, 666–671.

Amanatidou, A.; Slump, RA; Gorris, L.G.M.; Smid, E.J. High oxygen and high carbon dioxide modified atmospheres for shelf-life extension of minimally processed carrots. J. Foo Sci. 2000, 65, 61–66.

Ansah, F.A., Amodio, M.L., Colelli, G., 2018. Quality of fresh-cut products as affected by harvest and postharvest operations. J. Sci Food Agric. 98, 3614–3626.

Badiani, M., Fuhrer, J., Paolacci, A.R., Giovannozzi, S.G. Deriving Critical Levels for Ozone Effects on Peach Trees (Prunus persica L. Batsch) Grown in Open-top Chambers in Central Italy. Fresenius Environmental Bulletin 1996, 5, 592–597.

Barth, M.M., Zhou, C., Mercier, J., Payne, F.A. Ozone Storage Effects on Antocyanin Content and Fungal Growth in Blackberries. Journa of Food Science 1995, 60 (6), 1286–1288.

Baslam, M.; Morales, F.; Garmendia, I.; Goicoechea, N. Nutritional quality of outer and inner leaves of green and red pigmented lettuces (Lactuca sativa L.) consumed as salads. Sci. H rtic. 2013, 151, 103–111.

Baur, S., Klaiber, R.G., Koblo, A., Carle, R. Effect of Different Washing Procedures on Phenolic Metabolism of Shredded Packaged Iceberg Lettuce During Storage. Journa of Agricultural and Food Chemistry 2004, 52 (23), 7017–7025.

Baur, S.; Klaiber, R.; Hammes, W.P.; Carle, R. Sensory and microbiological quality of shredded, packaged iceberg lettuce as affected by pre-washing procedures with chlorinated and ozonated water. Innov. Food Sci. Emerg. Techno . 2004, 5, 45–55. [Cross ef].

Beltran, D., Selma, M.V., Marin, A., Gil, M.I. Ozonated Water Extends the Shelf Life of Fresh-cut Lettuce. Journa of Agricultural and Food Chemistry 2005, 53 (14), 5654–5663.

Bermudez-Aguirre, D., Barbosa-Canovas, G.V., 2013. Disinfection of selected vegetables under nonthermal treatments: chlorine, acid citric, ultraviolet light and Ozone. Food C ntrol 29, 82–90.

Bill, M., Sivakumar, D., Thompson, A.K., Korsten, L., 2014. Avocad fruit quality management during the postharvest supply chain. Food R v. Int. 30, 169–202.

Bridges, D.F.; Rane, B.; Wu, V.C.H. The effectiveness of closed-circulation gaseous chlorine dioxide or ozone treatment against bacterial pathogens on produce. Food C ntrol 2018, 91, 261–267.

Britton, H.C.; Draper, M.; Talmadge, J.E. Antimicrobial efficacy of aqueous Ozone in combination with short-chain fatty acid buffers. Infect Prev. ract. 2020, 2, 100032.

Buffle, M.-O.; Schumacher, J.; Meylan, S.; Jekel, M.; von Gunten, U. Ozone treatment and Advanced Oxidation of Wastewater: Effect of O3 Dose, pH, DOM and HO on Ozone decomposition and HO generation. Ozone ci. Eng. 2006, 28, 247–259.

Calatayud, A.; Barreno, E. Response to Ozone in two lettuce varieties on chlorophyll a fluorescence, photosynthetic pigments and lipid peroxidation. Plant hysiol. Bioche . 2004, 42, 549–555

Cantwell, M. Optimum Procedures for Ripening Tomatoes. Manag. Fruit ipening, Postharvest Hortic. Ser. 2000, 9, 80–88.

Chauhan, O.P.; Raju, P.S.; Ravi, N.; Singh, A.; Bawa, A.S. Effectiveness of Ozone in combination with controlled atmosphere on quality characteristics including lignification of carrot sticks. J. Foo Eng. 2011, 102, 43–48.

Cho, M.; Chung, H.; Yoon, J. Disinfection of water containing natural organic matter by using ozone-initiated radical reactions. Appl. nviron. Microb ol. 2003, 69, 2284–2291.

Das, E.; Gürakan, G.C.; Bayındırlı, A. Effect of controlled atmosphere storage, modified atmosphere packaging and gaseous ozone treatment on the survival of Salmonella Enteritidis on cherry tomatoes. Food M crobiol. 2006, 23, 430–438.

De Alencar, E.R., Faroni, A., Da Silva Pinto, M., Da Costa, A.R., Da Silva, T.A., 2013. Postharvest quality of ozonized" nanic? o" cv. Banana . Rev. C enc. Agron 44, 107.

Denis, N.; Zhang, H.; Leroux, A.; Trudel, R.; Bietlot, H. Prevalence and trends of bacterial contamination in fresh fruits and vegetables sold at retail in Canada. Food C ntrol 2016, 67, 225–234. [Cross ef].

Evrendilek, G.A. and Ozdemir, P., 2019. Effect of various forms of non-thermal treatment of the quality and safety in carrots. LWT, 105, 344–354.

Fan, X.; Sokorai, KJB; Engemann, J.; Gurtler, J.B.; Liu, Y. Inactivation of listeria innocua, Salmonella Typhimurium, and Escherichia coli O157:H7 on surface and stem scar areas of tomatoes using in-package ozone treatment. J. Foo Prot. 2012, 75, 1611–1618.

Fan, X.; Sokorai, KJB; Gurtler, J.B. Advanced oxidation process for the inactivation of Salmonella typhimurium on tomatoes by combination of gaseous Ozone and aerosolized hydrogen peroxide. Int. J Food Microbiol. 2020, 312, 108387.

Finch, G.R.; Black, E.K.; Labatiuk, C.W.; Gyurek, L.; Belosevic, M. Comparison of Giardia lamblia and Giardia muris cyst inactivation by Ozone. Appl. nviron. Microb ol. 1993, 59, 3674–3680.

Forney, C.F.; Song, J.; Hildebrand, P.D.; Fan, L.; McRae, K.B. Interactive effects of Ozone and 1-methylcyclopropene on decay resistance and quality of stored carrots. Postha vest Biol. Technol. 2007, 45, 341–348.

Galgano, F.; Caruso, M.C.; Condelli, N.; Stassano, S.; Favati, F. Application of Ozone in fresh-cut iceberg lettuce refrigeration. Adv. H rtic. Sci. 2015, 29, 61–64.

Garcia, A.; Mount, J.R.; Davidson, P.M. Ozone and Chlorine Treatment of Minimally Processed Lettuce. J. Foo Sci. 2003, 68, 2747–2751. [Cross ef].

García-Gimeno, R.M.; Zurera-Cosano, G. Determination of ready-to-eat vegetable salad shelf-life. Int. J. Food Microbiol. 1997, 36, 31–38.

Garg, N.; Churey, J.J.; Spittstoesser, D.F. Effect of Processing Conditions on the Microflora of Fresh-Cut Vegetables. J. Foo Prot. 1990, 53, 701–703.

Gibson, K.E.; Almeida, G.; Jones, S.L.; Wright, K.; Lee, J.A. Inactivation of bacteria on fresh produce by batch wash ozone sanitation. Food C ntrol 2019, 106, 106747.

Glassmeyer, S.T.; Shoemaker, JA Effects of Chlorination on the Persistence of Pharmaceuticals in the Environment. Bull. nviro . Contam Toxicol. 2005, 74, 24–31.

Gross, K.C.; Wang, C.Y.; Saltveit, M.E. Cold and Chilled Storage Technology; Dellino, C.V.J., Ed.; Springer US: Boston, MA, USA, 1997; ISBN 978-1-4612-8430-7.

Gu, G.; Bolten, S.; Mowery, J.; Luo, Y.; Gulbronson, C.; Nou, X. Susceptibility of foodborne pathogens to sanitizers in produce rinse water and potential induction of viable but non-culturable state. Food C ntrol 2020, 112, 107138. [Cross ef].

Gurol, M.D.; Singer, P.C. Kinetics of ozone decomposition: A dynamic approach. Enviro . Sci. T chnol. 1982, 16, 377–383.

Han, Y.; Floros, J.D.; Linton, R.H.; Nielsen, S.S.; Nelson, P.E. Response Surface Modeling for the Inactivation of Escherichia coli O157:H7 on Green Peppers (Capsicum annuum) by Ozone Gas Treatment. J. Foo Sci. 2002, 67, 1188–1193.

Hartley, L.; Igbinedion, E.; Holmes, J.; Flowers, N.; Thorogood, M.; Clarke, A.; Stranges, S.; Hooper, L.; Rees, K. Increased consumption of fruit and vegetables for the primary prevention of cardiovascular diseases. Cochra e Database Syst. Rev. 2013, 2013.

Haslay, C.; Leclerc, H. Microbiologie des eaux d'alimentation; Lavoisier TEC & DOC: Paris, France, 1993; ISBN 978-2-85206-918-3.

Hassenberg, K.; Idler, C.; Molloy, E.; Geyer, M.; Plöchl, M.; Barnes, J. Use of Ozone in a lettuce-washing process: An industrial trial. J. Sci Food Agric. 2007, 87, 914–919.

Hildebrand, P.D.; Forney, C.F.; Song, J.; Fan, L.; McRae, K.B. Effect of a continuous low ozone exposure (50 nL L−1) on decay and quality of stored carrots. Postha vest Biol. Technol. 2008, 49, 397–402.

Hillman, J.P.; Hill, J.; Morgan, J.E.; Wilkinson, J.M. Recycling of sewage sludge to grassland: A review of the legislation to control of the localization and accumulation of potential toxic metals in grazing systems. Grass orage Sci. 2003, 58, 101–111.

Hodges, R.J., Buzby, J.C., Bennett, B., 2011. Postha vest losses and waste in developed and less developed countries: opportunities to improve resource use. J. Agr Sci. 149, 37–45.

Hunt, N.K.; Mariñas, B.J. Inactivation of Escherichia coli with Ozone: Chemical and inactivation kinetics. Water es. 1999, 33, 2633–2641.

Huyskens-Keil, S., Hassenberg, K., Herppich, W., 2012. Impact of postharvest UV-C and ozone treatment on textural properties of white asparagus (Asparagus officinalis L.). J. App . Bot. Food Q al. 84, 229.

Ijaz, S., 2016. Molecu ar biology of ethylene, a review. J. Pla t. Mol. B eed .

Ishizaki, K.; Shinriki, N.; Matsuyama, H. Inactivation of Bacillus spores by gaseous Ozone. J. App . Bacteriol. 1986, 60, 67–72.

Jamil, A.; Farooq, S.; Hashmi, I. Ozone Disinfection Efficiency for Indicator Microorganisms at Different pH Values and Temperatures. Ozone ci. Eng. 2017, 39, 407–416.

Jouyban, Z., 2012. Ethyle e biosynthesis. Tech. Eng Appl. Sci 1, 107–110.

Kader, A.A., 2004. Increasing food availability by reducing postharvest losses of fresh produce. V Inte national Postharvest Symposium 682, 2169–2176.

Karaca, H.; Velioglu, Y.S. Effects of ozone treatments on microbial quality and some chemical properties of Lettuce, spinach, and parsley. Postha vest Biol. Technol. 2014, 88, 46–53.

Karaca, H.; Velioglu, Y.S. Ozone Applications in Fruit and Vegetable Processing. Food R v. Int. 2007, 23, 91–106.

Kim, J.-G.; Yousef, A.E.; Chism, G.W. Use of Ozone to inactivate microorganisms on Lettuce. J. Foo Saf. 1999, 19, 17–34.

Kleiber, T.; Borowiak, K.; Schroeter-Zakrzewska, A.; Budka, A.; Osiecki, S. Effect of ozone treatment and light colour on photosynthesis and yield of Lettuce. Sci. H rtic. 2017, 217, 130–136.

Kroupitski, Y.; Pinto, R.; Brandl, M.T.; Belausov, E.; Sela, S. Interactions of Salmonella enterica with lettuce leaves. J. App . Microbiol. 2009, 106, 1876–1885. [Cross ef].

Lewis, L., Zhuang, H., Payne, F.A., Barth, M.M. Beta-carotene Content and Color Assessment in Ozone-treated Broccoli Florets During Modified Atmosphere Packaging. In 199 IFT Annual Meeting Book of Abstracts. Instit te of Food Technologists: Chicago, 1996.

Liew, C.L.; Prange, R.K. Effect of Ozone and Storage Temperature on Postharvest Diseases and Physiology of Carrots (Caucus carota L.). Journal of the American Society for Horticultural Science 1994a, 119, 563–567.

Liew, C.L.; Prange, R.K. Effect of Ozone and Storage Temperature on Postharvest Diseases and Physiology of Carrots (Daucus carota L.). J. Am. Soc. H rtic Sci. 1994b, 119, 563–567.

Luwe, M.W.F., Takahama, U., Heber, U. Role of Ascorbate in Detoxifying Ozone in the Apoplast of Spinach (Spinacia oleracea L.) Leaves. Plant hysiology 1993, 101, 969–976.

Ma, L.; Zhang, M.; Bhandari, B.; Gao, Z. Recent developments in novel shelf life extension technologies of fresh-cut fruits and vegetables. Trends Food Sci. Technol. 2017, 64, 23–38. [Cross ef].

Mah, T.-F.C.; O'Toole, G.A. Mechanisms of biofilm resistance to antimicrobial agents. Trends Microbiol. 2001, 9, 34–39.

Marino, M.; Maifreni, M.; Baggio, A.; Innocente, N. Inactivation of Foodborne Bacteria Biofilms by Aqueous and Gaseous Ozone. Front. Microb ol. 2018, 9.

Minas, I.S., Vicente, A.R., Dhanapal, A.P., Manganaris, G.A., Goulas, V., Vasilakakis, M., Crisosto, C.H., Molassiotis, A., 2014. Ozone-induced kiwifruit ripening delay is mediated by ethylene biosynthesis inhibition and cell wall dismantling regulation. Plant ci. 229, 76–85.

Nadas, A., Olmo, M., Garcia, J., 2003. Growth of Botrytis cinerea and strawberry quality in ozone-enriched atmospheres. J. Foo Sci. 68, 1798–1802.

O'Donnell, C., Tiwari, B.K., Cullen, P.J. and Rice, R.G. eds., 2012. Ozone in food processing. John Wiley & Sons.

Ogden, M. Ozone treatment today. Ind. W ter ng. 1970, 7, 36–42.

Opara, UL, 2013. A revi w on the role of packaging in securing food system: adding value to food products and reducing losses and waste. Afr. J Agric. 8, 2621–2630.

Oztekin, S., Zorlugenc, B., Zorlugenc, F.K. Effects of Ozone Treatment on Microflora of Dried Figs. Journa of Food Engineering 2005, 75 (3), 396–399.

Pak, H., Dixon, J., 2001. Postha vest treatment with Ozone for the control of ripe rots in avocado. NewZea and Avocado Growers Association Annual Research Report. pp. 1.

Palou, L., Crisosto, C.H., Smilanick, J.L., Adaskaveg, J.E., Zoffoli, J.P. Effects of Continuous 0.3 ppm Ozone Exposure on Decay Development and Physiological Responses of Peaches and Table Grapes in Cold Storage. Postha vest Biology and Technology 2002, 24, 39–48.

Palou, L., Smilanick, J.L., Crisosto, C.H., Mansour, M. Effect of Gaseous Ozone Exposure on the Development of Green and Blue Molds on Cold Stored Citrus Fruit. Plant isease 2001, 85 (6), 632–638.

Pan, G.Y.; Chen, C.-L.; Chang, H.-M.; Gratzl, J.S. Studies on Ozone Bleaching. I. The Ef ect of PH, Temperature, Buffer Systems and Heavy Metal-Ions on Stability of Ozone in Aqueous Solution. J. Woo Chem. Technol. 1984, 4, 367–387.

Patil, S.; Bourke, P.; Frias, J.M.; Tiwari, B.K.; Cullen, P.J. Inactivation of Escherichia coli in orange juice using Ozone. Innov. Food Sci. Emerg. Techno . 2009, 10, 551–557.

Patil, S.; Valdramidis, V.P.; Cullen, P.J.; Frias, J.; Bourke, P. Inactivation of Escherichia coli by ozone treatment of apple juice at different pH levels. Food M crobiol. 2010, 27, 835–840.

Perez, A.G., Sanz, C., Rios, J.J., Olias, R., Olias, J.M. Effects of Ozone Treatment on Postharvest Strawberry Quality. Journa of Agricultural and Food Chemistry 1999, 47, 1652–1656.

Ranieri, A., D'Urso, G., Nali, C., Lorenzini, G., Soldatini, G.F. Ozone Stimulates Apoplastic Antioxidant Systems in Pumpkin Leaves. Physio ogia Plantarum 1996, 97, 381–387.

Restaino, L.; Frampton, E.W.; Hemphill, J.B.; Palnikar, P. Efficacy of ozonated water against various food-related microorganisms. Appl. nviron. Microb ol. 1995, 61, 3471–3475.

Rice, RG; Robson, C.M.; Miller, G.W.; Hill, AG Uses of Ozone in drinking water treatment. J. Am. Water orks Assoc. 1981, 73, 44–57

Rodoni, L., Casadei, N., Concellon, A., Chaves Alicia, A.R., Vicente, A.R., 2009. Effect of short-term ozone treatments on tomato (Solanum lycopersicum L.) fruit quality and cell wall degradation. J. Agr c. Food C em. 58, 594–599.

Sachadyn-Krol, M., Materska, M., Chilczuk, B., Karaś, M., Jakubczyk, A., Perucka, I., Jackowska, I., 2016. Ozone-induced changes in the content of bioactive compounds and enzyme activity during storage of pepper fruits. Food C em. 211, 59–67.

Salehi, B.; Sharifi-Rad, R.; Sharopov, F.; Namiesnik, J.; Roointan, A.; Kamle, M.; Kumar, P.; Martins, N.; Sharifi-Rad, J. Beneficial effects and potential risks of tomato consumption for human health: An overview. Nutrit on 2019, 62, 201–208.

Sarig, P., Zahavi, T., Zutkhi, Y., Yannai, S., Lisker, N., Ben-Arie, R. Ozone for Control of Posthar-vest Decay of Table Grapes Caused by Rhizopus stolonifer. Physio ogical and Molecular Plant Pathology 1996, 48, 403–415.

Sarron E., Gadonna-Widehem P., Aussenac T., 2021. Ozone treatments for preserving fresh vegeta-bles quality: A critical review. Foods 10(3):605

Seo, K.H.; Frank, J.F. Attachment of Escherichia coli O157:H7 to Lettuce Leaf Surface and Bacterial Viability in Response to Chlorine Treatment as Demonstrated by Using Confocal Scanning Laser Microscopy. J. Foo Prot. 1999, 62, 3–9.

Sharma, K.D.; Karki, S.; Thakur, N.S.; Attri, S. Chemical composition, functional properties and processing of carrot—a review. J. Foo Sci. Technol. 2012, 49, 22–32.

Sharpe, D.; Fan, L.; McRae, K.; Walker, B.; MacKay, R.; Doucette, C. Effects of Ozone Treatment on Botrytis cinerea and Sclerotinia sclerotiorum in Relation to Horticultural Product Quality. J. Foo Sci. 2009, 74, M250–M257.

Shen, C.; Norris, P.; Williams, O.; Hagan, S.; Li, K. Generation of chlorine byproducts in simulated wash water. Food C em. 2016, 190, 97–102. [Cross ef].

Singh, N.; Singh, R.K.; Bhunia, A.K.; Stroshine, R.L. Efficacy of Chlorine Dioxide, Ozone, and Thyme Essential Oil or a Sequential Washing in Killing Escherichia coli O157:H7 on Lettuce and Baby Carrots. LWT Fo d Sci. Technol. 2002, 35, 720–729.

Skog, L.J., Chu, C.L. Effect of Ozone on Qualities of Fruits and Vegetables in Cold Storage. Canadi n Journal of Plant Science 2001, 81 (4), 773–778.

Smilanick, J.L., 2003. Use of Ozone in storage and packing facilities. Washin ton Tree Fruit Posthar-vest Conference 1–10.

Smith, S.M.; Scott, J.W.; Bartz, J.A.; Sargent, S.A. Effect of Time After Harvest on Stem Scar Water Absorption in Tomato. HortSc ence 2007, 42, 1227–1230.

Toti, M., Carboni, C., Botondi, R., 2018. Postharvest gaseous ozone treatment enhances quality parameters and delays softening in cantaloupe melon during storage at 6° C. J. J. Environ. Sci. H alth B 98, 487–494.

Tran, T., Aimla-Or, S., Srilaong, V., Jitareerat, P., Wongs-Aree, C., Uthairatanakij, A., 2013. Fumi-gation with Ozone to extend the storage life of mango fruit cv. Nam Dok Mai No. 4. Agric. Food Sci. J. Ghana 44, 663–672.

Tzortzakis, N.; Borland, A.; Singleton, I.; Barnes, J. Impact of atmospheric ozone-enrichment on quality-related attributes of tomato fruit. Postha vest Biol. Technol. 2007a, 45, 317–325.

Tzortzakis, N.; Singleton, I.; Barnes, J. Deployment of low-level ozone-enrichment for the preser-vation of chilled fresh produce. Postha vest Biol. Technol. 2007b, 43, 261–270.

Wang, K.L.C., Li, H., Ecker, J.R., 2002. Ethylene biosynthesis and signaling networks. Plant Cell 14, S131–S151.

Wang, L.; Fan, X.; Sokorai, K.; Sites, J. Quality deterioration of grape tomato fruit during stor-age after treatments with gaseous Ozone at conditions that significantly reduced populations of Salmonella on stem scar and smooth surface. Food C ntrol 2019, 103, 9–20.

Wani, S.; Barnes, J.; Singleton, I. Investigation of potential reasons for bacterial survival on 'ready-to-eat' leafy produce during exposure to gaseous Ozone. Postha vest Biol. Technol. 2016, 111, 185–190.

Whangchai, K., Saengnil, K., Uthaibutra, J., 2006. Effect of Ozone in combination with some organic acids on the control of postharvest decay and pericarp browning of longan fruit. Crop P ot. 25, 821–825.

WHO. Develo ment of WHO Nutrition Guidelines; WHO: Geneva, Switzerland, 2018.

Yahia, E.M. ed., 2019. Postharvest technology of perishable horticultural commodities. Woodhead Publishing.

Yang, Y.; Komaki, Y.; Kimura, SY; Hu, H.-Y.; Wagner, E.D.; Mariñas, B.J.; Plewa, M.J. Toxic Impact of Bromide and Iodide on Drinking Water Disinfected with Chlorine or Chloramines. Enviro . Sci. T chnol. 2014, 48, 12362–12369. [Cross ef]

Yaseen, T., Ricelli, A., Turan, B., Albanese, P., D'onghia, A.M., 2015. Ozone for postharvest treatment of apple fruits. Phytop thol. Mediterr. 54, 94–103.

Yin, X.R., Zhang, Y., Zhang, B., Yang, S.L., Shi, Y.N., Ferguson, I.B., Chen, K.S., 2013. Effect of acetylsalicylic acid on kiwifruit ethylene biosynthesis and signaling components. Postha vest Biol. Technol. 83, 27–33.

Zambre, S.S.; Venkatesh, K.V.; Shah, N.G. Tomato redness for assessing ozone treatment to extend the shelf life. J. Foo Eng. 2010, 96, 463–468.

Zhang, L., Lu, Z., Yu, Z. and Gao, X. Preservation Fresh-cut Celery by Treatment of Ozonated Water. Food C ntrol 2005, 16, 279–28.

Ozone Applications in Meat Processing and Seafood

4.1   Introduction

Over the years, the food business has developed several cutting-edge technologies for food preservation, including high hydrostatic pressure, radiofrequency, high-intensity pulsed electric fields, ultrasound, irradiation, and ozone treatment. Each of these systems has benefits, drawbacks, and restrictions based on several variables, including the kind of food, temperature, pH, the existence of microbes, and national laws. Due to customer demands for fresh, safe products with minimal processing, unaffected nutritional qualities, and no chemical residues following treatment, interest in ozone has recently rekindled (Gimenez et al. 2021; Xue et al. 2023). Ozonation is a cost-effective and environmentally friendly technology that can be used in either liquid or gaseous food. Numerous methods can be employed on-site to produce ozone; currently, the most widely utilized ones for commercial purposes are electrolysis, UV radiation, and corona discharge (Prabha et al. 2015; Gimenez et al. 2021; Xue et al. 2023).

Additionally, the synthesis of ozone on-site reduces the need for storage and shipping. One potent oxidant that can be used in the food business is ozone. It is an allotropic form of oxygen that acts against various foodborne pathogens and spoiling organisms. It is considered a broad-spectrum antimicrobial agent with more potent antibacterial activity than chlorine (Priyanka et al. 2014). Ozone has demonstrated efficacy against both Gram-positive and Gram-negative bacteria.

Moreover, it can degrade mycotoxins present in fruits, vegetables, meat, grains, and their derivatives, as well as inactivate viruses and fungi (Premjit et al. 2014; Brodowska et al. 2017; Pandiselvam et al. 2018; Niveditha et al. 2021). Ozone has been utilized since the nineteenth century for various purposes, including air sterilization, crop spraying without hazardous chemicals, deodorizing industrial waste, washing and disinfecting

J. A. Parray et al., *Ozone Technology for Food Processing and Preservation*, Synthesis Lectures on Chemical Engineering and Biochemical Engineering, https://doi.org/10.1007/978-3-031-81461-7_4

equipment, and treating water. When food products are present, excess ozone autodecomposes quickly to create oxygen and leaves no trace (Oner et al. 2011; Pandiselvam et al. 2018). The Food and Drug Administration (FDA) officially approved Ozone in 2001 for use in the food industry and direct contact with food products, such as fish, beef, and poultry. In 1997, ozone was classified as GRAS (Generally Recognised as Safe) (Fig. 4.1). Kim and associates (1999), Gonçalves (2009).

Because ozone has a high potential for oxidation and reduction, it first oxidizes the components of microbial cell walls before entering the microorganisms. Once inside, it also oxidizes vital components like proteins, enzymes, unsaturated lipids, and nucleic acids. This damage to the cell wall and membrane results in the death of the bacterial cells (Greene et al. 2012; Brodowska et al. 2017; Pandiselvam et al. 2017). The literature claims that ozone causes critical cellular components in microorganisms to oxidize gradually. Victorin (1992) discovered two ways that ozone kills microorganisms: (a) it oxidizes the amino acids and sulfhydryl groups in proteins, enzymes, and peptides to produce shorter peptides; (b) it oxidizes the double bonds in polyunsaturated fatty acids. Cell lysis is the outcome of the unsaturated lipids breaking down. According to Kim et al. (1999), ozone causes bacterial cell permeability and subsequent lysis in Gram-negative bacteria by first

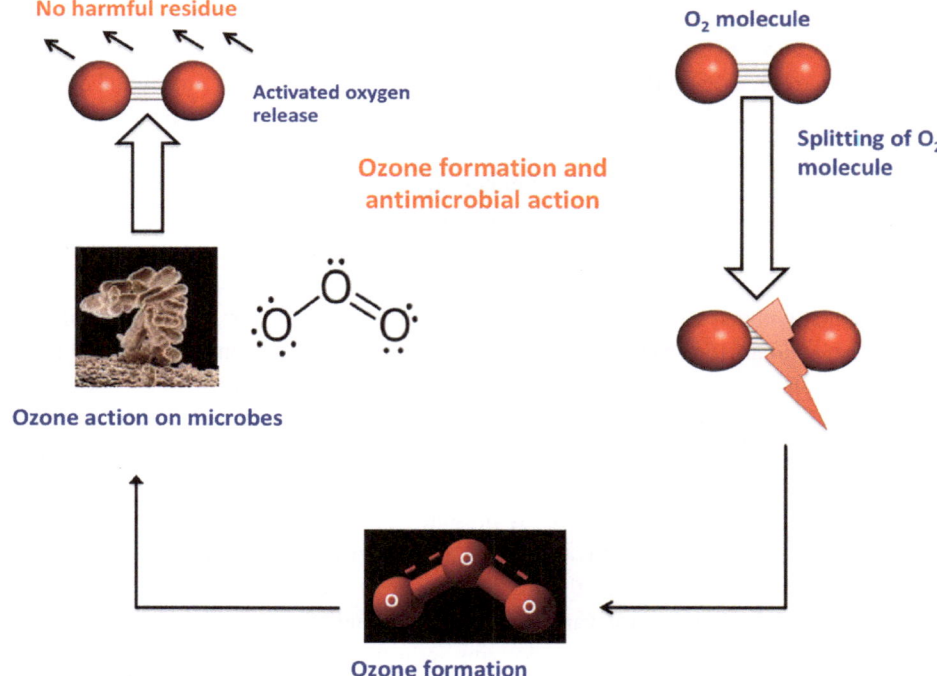

Fig. 4.1 An overview of ozone generation and bacterial killing

attacking the lipoprotein and lipopolysaccharide layers. Because peptidoglycan is present in the walls of gram-positive bacteria, they are more resistant than gram-negative bacteria (Pandiselvam et al. 2022b). Salmonella, E. Coli, Shigella, and other human infections are within the varied group of bacteria known as Enterobacteriaceae; E. Coli is the Gram-negative bacteria most frequently investigated (Khadre et al. 2001). According to Ingram and Haines (2009), ozone disrupts E. Coli's respiratory system, ultimately leading to its demise.

The concentration, temperature, and ozone application mode all affect how effective the microbicidal action is. Temperature is one of the most crucial variables since it influences the gas's stability, reactivity, and solubility. Ozone can be added to food products as a gas or dissolved in water (Khadre et al. 2001; Coll Cardenas et al. 2011). Other elements influencing its performance include the amount of organic matter around the cells and the intrinsic qualities of the meal, such as pH, aw, and additives (Manousaridis et al. 2005; Priyanka et al. 2014). A flow chart summarising the effectiveness factors of ozone, decontamination methods against mold, bacteria, fungi, and biofilms, and the use of ozone in conjunction with other preservation technologies (hurdle technology) was presented by Xue et al. (2023). Listeria monocytogenes was the pathogen (together with Salmonella typhimurium, Yersinia enterocolitica, and Staphylococcus aureus) that was determined to be most sensitive to ozone by Restaino et al. (1995). For ozone to be used safely and effectively, precise treatment requirements for each product must be established. Due to the quick breakdown of organic compounds, ozone also has the ability to absorb flavors and odd smells in the water.

Similarly, ozone has a role in deodorizing the air (Gonçalves 2009). This mini-review aims to examine and condense the impact of ozonation treatments, both liquid and gaseous, on various kinds of meat and meat products. To compare the results reported by multiple authors, emphasis is put on the ozone concentrations employed in the different treatments and the units used to express these concentrations in liquid or gaseous phases. The impact of ozone treatment on the physicochemical properties of meat products is also covered in detail in this chapter.

4.2 Contamination Risks in Meat Process and Economic Losses

A wide range of microflora, including molds, yeasts, and bacteria—some of which are pathogens—grow beautifully on meat and meat products. The animals' skin and digestive tracts are the primary suppliers of these microorganisms, sometimes known as pathogens. These pathogens' composition mainly depends on pre-slaughter monitoring procedures, animal age, handling during slaughter, temperature management, preservation techniques, and consumer handling (Saad et al. 2019; Shaltout et al. 2018). Meat contains a variety of bacterial, mold, and yeast species that existed prior to the meat rotting process. Numerous bacterial species, including Streptococcus, Micrococcus, Pseudomonas, Bacillus, and

Clostridium spp., are found in meat and meat products, along with yeast species like Cryptococcus and Candida spp. and mold species, including Cladosporium, Geotrichum, and Penicillin spp. Enterococcus is the most prevalent species of bacteria (Shaltout et al. 2017a, 2017b). Different species of bacteria can harm meat and meat products under various storage conditions. Enteric bacteria are found on products that have been chilled, and certain species of bacteria are more productive in colder environments than in hotter ones.

Similarly, certain species, such as Acinetobacter and Moraxella, are likelier to flourish in hotter climates. Because they are salt-resistant, many viruses prone to proliferate in a hotter climate are usually found in raw, salted, cured items like hams and uncooked beef (Shaltout and Abdel Aziz 2004). Pseudomonas species and enteric bacteria grow best in environments with a temperature of Celsius, and they are most common in cold environments with five modifications. Meat shelf life is just two degrees Celsius due to the Pseudomonas bacterial species, which can alter the temperature and develop considerably slower. Similar to this, several other species, such as Salmonella spp. Above 0 °C can also affect the quality and shelf life of meat. However, under 7 °C, the growth of Salmonella spp. is too slow, and a pH of 5.5–7.0 is ideal for developing spoilage bacteria. Within this range, bacterial activity leads to the formation of slime, a foul odor in the meat, and changes in the appearance of the meat. Several ammonia derivatives, including methylamine and others, were found in the spoiling activity. Various ketones and alcohols were also formed by the bacterial activity with a pleasant smell (Shaltout et al. 2016).

4.2.1 The Lipid Oxidation of Meat Spoilage

Meat oxidation and free radical generation will impact the meat's fatty acids, leading to an unpleasant odor, a sour taste, and a decline in meat quality. Following the animal's slaughter, the fatty acids in the meat begin to oxidize, which stops the metabolic process and blood circulation. The lipid oxidation process has three main steps: initiation, propagation, and termination. During this process, the fatty acid constantly forms a double bond with the oxygen in the air (Shaltout and Hashim 2002).

4.2.1.1 The Initiation of Meat Spoilage
Throughout the process, natural catalysts such as heat and radiation create free radicals, which combine with oxygen in the air to make peroxyl radicals (Shaltout et al. 2013). The propagation of meat spoilage occurs when newly generated free radicals and hydroperoxides are formed by the reaction of peroxyl radicles formed during the initiation step with other lipid molecules.

4.2.1.2 The Termination of Meat Spoilage
Meat spoilage occurs when the free radicles produced in the first two steps interact and form non-radicle products. The different factors that affect the oxidation of the lipids are

the antioxidant vitamin E and the composition of the fatty acid. Hydroperoxide breakdown causes the release of several products, such as acids, ketones, and aldehydes. This effect causes nutritional value degradation and loss of color because of their extreme impact on carbohydrates, lipids, and vitamins. Ultimately, they are related to several extreme pathogenic processes (Shaltout et al. 2020). There are two ways to hydrolyze the lipid in meat: enzymatically and non-enzymatically. Enzymatic hydrolysis involves the assistance of multiple enzymes, including lipase and phospholipase, but Phospholipase A1 and A2 are the primary enzymes involved in the hydrolysis of the lipid in the meat. Haemoglobin, cytochrome, and myoglobin are the three proteins that cause non-enzymatic hydrolysis, are oxidatively sensitive, and generate hydroperoxides (Shaltout 2023).

4.2.1.3 The Autolytic Enzymatic Activity of Meat Spoilage

The body's enzymes are the primary factor contributing to the degradation of meat. They only usually function when an animal is alive; if they pass away or are killed, they are the primary source of meat deterioration and shortened shelf life. The enzymes function as a catalyst for chemical reactions with other chemicals, which initiates the decline of meat (Hassan and Shaltout 2004). The enzyme tissue protease, released during the breakdown of polypeptides, is in charge of the flavor and textural changes. Meat tenderisation occurs at low temperatures, where biogenic amines and bacteria production increases. This increases the proteolytic enzymes that deteriorate meat quality (Shaltout et al. 2016). Some enzymes, such as calpains and cathepsins, catalyze the post-mortem autolysis of the meat through the digestive process (Farag et al. 2023).

4.3   Sustainable Solutions to Meat Processing and Preservation

4.3.1   Gaseous Ozone Treatments

Ozone can be applied in gaseous or dissolved in aqueous phase in the food industry. The various units of measurement used by the authors when reporting ozone concentrations in the literature must be considered. When applying ozone in the liquid phase, concentrations can be expressed as ppm or mg/L. When treating the gaseous phase, concentrations in the air can be represented by volume or weight. When volumetric concentrations are used, the equivalencies are as follows: 1 g O_3/m^3 = 467 ppmv O_3; 1 ppmv O_3 = 2.14 mg O_3/m^3. When concentrations of ozone in the air by weight, the corresponding equivalencies are as follows: 100 g O_3/m^3 = 7.8% O_3; 1% O_3 = 12.8 g O_3/m^3; 1% O_3 = 7,284 ppm Ozone. Since ozone is an unstable gas that cannot be stored, it must be produced locally whenever required. The concentration needed determines the ozone-generating processes. The UV photochemical method produces ozone at low concentrations of up to 0.3–0.4% by weight due to the air being exposed to radiation. Feed gas, typically ambient air,

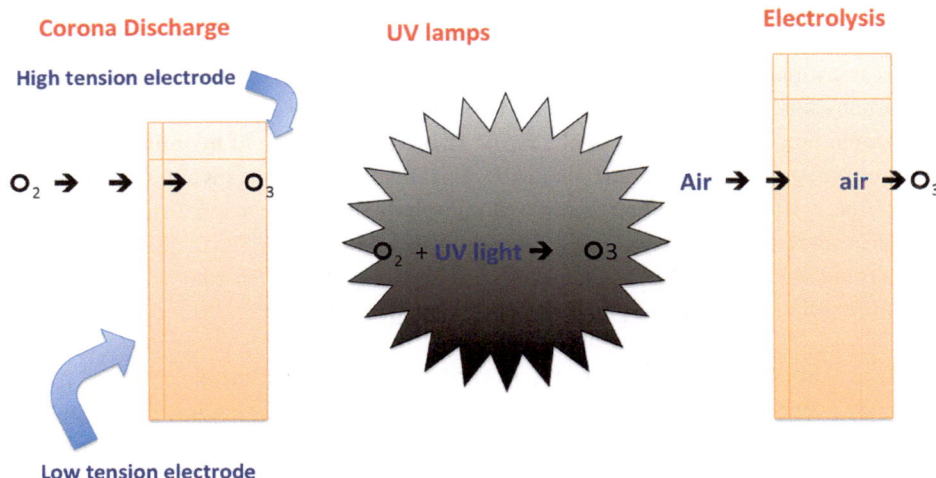

Fig. 4.2 Ozone treatment process in prevention of meat spoilage

is passed through a UV lamp (185 nm in wavelength). Photo disassociation splits the oxygen molecules into unstable oxygen radical atoms, reacting with oxygen molecules to form ozone. The corona discharge process produces higher ozone concentrations, also known as plasma (Cameron and Rice 2012). This process involves passing gas—either air or dry oxygen—through electrodes spaced apart by a dielectric material. As the oxygen molecules move through the medium, they split into highly energetic radical atoms, or oxygen radicals, which combine with molecular oxygen to form ozone (Priyanka et al. 2014). Corona discharge is the most widely used commercial ozone production method (Fig. 4.2).

After ozone treatment, the extra ozone should be removed for safety reasons (Brodowska et al. 2017). Increased concentrations of ozone lead to more rapid Inactivation of microorganisms, resulting in shorter treatment periods and more miniature Decimal Reduction Time (D) values (Steenstrup and Floros 2004). Some food compounds may oxidize when exposed to higher ozone concentrations (Priyanka et al. 2014). Even brief exposure to high concentrations of ozone is harmful to human health. The destructive qualities of ozone may induce specific symptoms, such as drying of the throat, headache, irritation to the nose, possibly severe disease, and even death (Muthukumarappan et al. 2000). Long-term exposure to ozone is linked to a higher risk of respiratory illnesses, metabolic diseases, nervous system problems, poor birth outcomes, low male and female fertility in the reproductive system, cancer, and increased cardiovascular mortality. Ozone concentrations of 1.0–2.0 ppm cause irritation to the upper part of the throat, headache, chest pain, coughing, and dry throat after only a few minutes of exposure; concentrations of 5.0–10.0 ppm cause an elevated pulse and lung edema; concentrations over 50.0 ppm are potentially fatal; and concentrations over 1700 ppm are lethal (Brodowska et al. 2017).

According to Gonçalves (2009), the Environmental Protection Agency (EPA) set a limit concentration of 0.08 parts per million (ppm) for human exposure to ozone for eight hours. Furthermore, employees must be aware that many organic products can become significantly more flammable when producing ozone using oxygen as the feed gas (Brodowska et al. 2017). Ozone can be made in the aqueous phase electrolytically or by bubbling the gas through water to facilitate dissolution. The ozone solubility in water, which depends on the water's temperature, pressure, ionic strength, presence of ionic salts, and ozone gas concentration, must be considered in the event of gas bubbles. Poor ozone solubility lowers the concentration levels that can be achieved in aqueous solutions, which is one of the decisive elements that affect decontamination efficiency (Batagoda et al. 2018; Aslam et al. 2020). According to Brodowska et al. (2017), ozone is ten times more soluble in water than oxygen, and its solubility declines as water temperature rises. At 0 °C, ozone concentration in water is higher (0.6401 ozone/L water) than at higher temperatures. When the pH falls below 7.0, the gas dissolves in water; nonetheless, when the pH rises, ozone spontaneously breaks down, releasing highly reactive free radicals like hydroxyl_OH. Within ten minutes, at pH = 8, almost half of the added ozone breaks down into different intermediate forms and oxygen. Ozone decomposes in solution following a stepwise mechanism, producing hydroperoxyl, hydroxyl, and superoxide radicals. The hydroxyl radical is a critical transient species and chain-propagating radical. The reactivity of ozone is attributed to the tremendous oxidizing power of these free radicals (Brodowska et al. 2017). Ozone causes the formation of free radicals at pH > 8, and at lower pH, the mechanism of ionic reaction predominates (ozonolysis) and generates peroxide production (Gonçalves 2009). In some cases, gaseous ozone offers advantages over aqueous ozone due to its superior penetration capacity, enabling it to reach inaccessible areas in products where pathogens may be present (Shynkaryk et al. 2015).

4.3.2 Gaseous Ozone Treatments

Because meat provides the nutrients necessary for a balanced diet, people worldwide eat much meat (beef, chicken, pork, shellfish, etc.). Either fresh meat or processed meat is consumed. Meats are one of the primary sources of foodborne illness because of their high aw and nutrient content, which makes them vulnerable to microbial attack and shortens their shelf life (Fearnley et al. 2011; Antunes et al. 2016). Numerous factors can lead to contamination during the meat processing process, including insufficient sanitization of equipment, handling of equipment carelessly, evisceration and slaughtering practices, contaminated washing water, and unsuitable temperature settings (Pandiselvam et al. 2022a). Various elements influence ozone disinfection rates and vary depending on the type of organism. Epelle et al. (2023) divided these variables into environmental influences, substrate/material qualities, and operational properties. These variables may change the ozone's stability in the medium (water or air), the effectiveness of microbial Inactivation, or both. The effects of gaseous ozone treatment on several microorganisms found in diverse food matrices (beef, chicken, poultry, and seafood) are summarized in Table 4.1.

These effects vary depending on the application forms and the ozone concentrations employed. Because the authors frequently used ppm without clarifying whether these units are volumetric or weight-based, concentrations in the Table were expressed in the units provided to prevent errors. The largest microbial Inactivation was seen in beef samples treated with gaseous ozone after one day at 0 °C, according to Coll Cardenas et al. (2011). This resulted in a drop of 2.0 log10 cycles in the total aerobic mesophilic heterotrophic microbe counts and 0.7 Log cycles in the E. coli counts. Unfortunately, these treatments' lipid oxidation and surface color outcomes were undesirable. On the other hand, three hours of exposure to a gaseous ozone concentration of 154 mg/m^3 at either 0 °C or 4 °C reduced the counts of all aerobic mesophilic heterotrophic microorganisms by 0.5 log cycles and the counts of E. coli by 0.6–1.0 log cycles, without causing rancidity or altering the color of the beef. Ozone therapy enhanced CFU reduction and extended product shelf life when combined with cooling. The effects of ozone on ground Hanwoo beef infected with E. coli were investigated by Cho et al. (2014a). The treatment involved subjecting the infected samples to 10 mg O$_3$/h for three days at 4 °C in a 25 × 20 × 20 cm chamber. The results showed that after one day of ozone exposure, E. coli counts decreased by 0.53 log CFU/g, and no bacterial growth was seen over the three days of storage. The combined impact of ozone pretreatment and carbon monoxide (CO) on the vacuum-packed beef quality was examined by Lyu et al. (2016). Before vacuum packaging, beef samples were processed under MAP conditions for 1.5 h using gaseous mixtures of various volume ratios of ozone and carbon monoxide (100% CO; 2% O$_3$/ 98% CO; 5% O$_3$/95% CO; 10% O$_3$/90% CO). After forty-five days of storage at 0° C, the samples were assessed, and the combined pretreatment decreased total viable counts. Gimenez et al. (2021) used concentrations of 280 mg O$_3$/m^3 to treat beef samples with ozone pulses ranging from 5 to 10 min every 30 min for 5 h. This resulted in a reduction of more than 1 log in the counts of mesophilic Enterobacteriaceae and lactic acid bacteria, as well as a decrease in the counts of inoculated L. monocytogenes (102 CFU/ g tissue) to values below the detection limit. The samples were then refrigerated for 16 days at 4 °C. In 2010, Pirani researched the application of low-concentration gaseous ozone to mitigate or prevent the emergence of non-acceptable grey-black patches on the surface of fermented sausages, which are caused by heterogeneous molds. The ozone concentration in the treatment rooms was kept at 0.5 ppm throughout the trials. An eight-hour-per-day ozone treatment was carried out throughout the four-month ripening phase. The starter culture, P. Sangiovese, was able to proliferate, whereas the applied therapy suppressed the abnormal mold strains. Muthukumar and Muthuchamy (2013) employed 25 g of fresh chicken samples dipped in deionized water mixed with roughly 108 CFU/

Table 4.1 Effect of gaseous ozone concentration on meats

Type of meat	Gaseous ozone treatment	Tested Microorganism	Results	References
Chicken breasts	Gaseous ozone, > 2000 ppm (4.28 × 103 mg O_3/m^3) for up to 30 min or 15 min, followed by storage under 70% CO_2:30% N_2 (MAP). Storage temperature = 7 °C	Pseudomona aeruginosa and Salmonella infants	Reduction of 95% of P. aeruginosa and 97% of S. infantis counts immediately after ozone treatment Indigenous coliforms were unaffected MAP has little further impact. Shelflife and sensory aspects remained acceptable throughout the storage period of 9 days	Al-Haddad et al. (2005)
Fermented sausages	Ozone concentration was maintained at 0.5 ppm. The ozone treatment was conducted eight hours per day for four month	Heterogeneous molds	The applied treatment inhibited the growth of anomalous mold strains, and allowed the growth of the starter culture used, P. Sangiovese	Pirani (2010)
Beef	Gaseous ozone (154 × 10^{-6} kg m^{-3}) in culture media inoculated with *E. coli* after 3- or 24-h treatment at 0° and 4 °C	*E. coli*	Total Inactivation of microorganisms	Cardenas et al. (2011)
Fresh chicken	Gaseous ozone flow of 33 mg/min for 9 min	L. monocytogenes	Immediately after treatment, a significant decrease (4 log cycles) of L. monocytogenes counts	Muthukumar and Muthuchamy (2013)

(continued)

Table 4.1 (continued)

Type of meat	Gaseous ozone treatment	Tested Microorganism	Results	References
Fresh chicken legs	Ozone doses 2, 5, and 10 mg/L (2 × 103, 5 × 103 and 10 × 104 mg O_3/m^3) and vacuum packaging stored at 4 ± 1 C, for 16 days	Total viable counts Pseudomonas spp., LAB, yeasts and molds, and Enterobacteriaceae	Combination of gaseous ozone (5 and 10 mg/L) and vacuum packaging showed microbial counts < 7 CFU/ gduring 16 days, extending the shelflife for six days compared with the control	Gertzou et al. (2017)
Turkey Breast Meat	Ozone treatment: 1 × 104 mg/m^3, for up to 8 h	Counts of total aerobic mesophilicbacteria, Enterobacteriaceae, and yeast mold	Approximately 2.9, 2.3, and 1.9 log reductions were achieved in totalaerobic mesophilic bacteria, Enterobacteriaceae and yeast-mold respectively immediately after treatment	Ayranci et al. (2020)
Salmon	Ozone doses: 1 mg/ m^3 or 3 mg/m^3 Exposure times: 5 or 10 min	Photobacterium	The more extended treatments showed the largest decrease in microorganism counts (1–1.5 cycles log)	Qian et al. (2022)

mL of L. monocytogenes for 30, 45, and 60 s. The samples were then ozonated for 1–9 min at a dosage of 33 mg/min after being air-dried for 1 h in a laminar flow hood. The amount of L. monocytogenes that survived each ozonation cycle on the chicken was measured and contrasted with the non-ozonated samples. The study found that 2×106 CFU/g of L. monocytogenes could be effectively inactivated on chicken samples using ozone at 33 mg/min dosages for 9 min in the gaseous phase. Cho et al. (2014b) inoculated samples of chicken breast with S. Typhimurium and found that samples exposed to gaseous ozone (10 mg O_3/h in a chamber measuring $25 \times 20 \times 20$ cm) for one day reduced the number of colony-forming units (CFU/g) by 0.4 compared to the untreated inoculated meat (7.84 log CFU/g tissue), demonstrating the bacteriostatic effect of ozone. Following three days in storage, the ozone-treated samples showed a count of 7.51 CFU/g, while the control samples showed several 8.30 CFU/g. In contrast to single vacuum packaging, Gertzou et al. (2017) discovered that the combination of gaseous ozone (2, 5, and 10 mg/L) with vacuum packaging prolonged the shelf life of chicken legs under refrigeration for six days (5 and 10 mg/L). After ten days of storage, the counts of Pseudomonas, total viable counts (TVC), Enterobacteriaceae, and lactic acid bacteria (LAB) in fresh beef exceeded 7 log CFU/g tissue; in contrast, samples treated with ozone (5 and 10 mg/L) remained below this value for sixteen days. Commercial samples of pork meat were treated with ozone by Jaksch et al. (2004) to see if this would inhibit microbial growth and increase the shelf life of the meat products. The study employed Proton-Transfer-Reaction Mass Spectrometry (PTR-MS) to examine volatile emissions. The mass 63 signal, believed to represent dimethylsulphide, was identified and used as a marker for bacterial activity. The impact of combining vacuum cooling with an ozone-based re-pressurization technique (Invac) on Clostridium perfringens (G+) was investigated by Liao et al. in 2021. This treatment (150 mg O_3/m^3 for 30 min) doubled the shelf life of cooked pork by extending its dormant phase and lowering growth rates. After four days, control samples had surpassed 7 log CFU/g tissue, whereas samples treated with ozone had counts below 7 log CFU/g tissue for seven days. Ayranci et al. (2020) investigated the impact of gaseous ozone treatment on total aerobic mesophilic bacteria counts in turkey flesh samples at concentrations of 10 g O_3/m^3 and exposure durations of 2, 4, 6, and 8 h. The initial counts of mesophilic bacteria were found to be greatly reduced by all ozone treatments; values obtained ranged from 1.5 to 3 log decreases. In terms of enterobacteria, samples exposed for 2–4 h showed a drop in microbial counts of approximately 1–1.5 log units, while after 6 h, samples showed a decline of 2.3 log units. Compared to other meats, fresh fish and marine items are highly perishable. Such food quickly becomes less sanitary due to microbial cross-contamination from several sources, which eventually causes spoiling (Manousaridis et al. 2005). When fish were exposed to an initial ozone treatment (60 min) and a daily exposure (30 min) at concentrations of 270 mg O_3/m^3, the authors Da Silva et al. (1998) found a decrease of 1.0 log CFU/cm^2. The fish species studied were Pseudomonas putida, Shewanella putrefaciens, Brochothrix thermosphacta, Enterobacter sp., and Lactobacillus plantarum. Aponte et al. (2018) investigated the impact of six cycles of five minutes of eight parts per million (ppm) ozone on Enterobacteriaceae and Aeromonas spp. found in various fresh fish products (musk octopus and blanched fish) during days 0, 2, 5, 7, 9, and 12 of storage. With a drop of about 2 log CFU/g in ozonated musk octopus and less than 4 log CFU/g in ozonised scalded fish, ozonation demonstrated its effectiveness. The impact of gaseous ozone on the microbiological growth of salmon was investigated by Qian et al. (2022) at several doses and exposure durations (1 mg/m^3 or 3 mg/m^3 for 5 min and 1 or 3 mg/m^3 for 10 min). The microbe counts (1–1.5 cycles log) decreased more in the longer treatments.

4.3.3   Aqueous Ozone Treatments

Aqueous ozone treatment has become more critical in the food business because of its many advantages and wide range of uses. Aqueous ozone significantly increases food safety and acts as an efficient disinfectant by lowering bacteria, viruses, pathogens, and other microbes in meat products. Reagan et al. (1996) examined the trimming and washing of beef carcasses to improve the microbiological quality of meat. They compared treatments using hydrogen peroxide or ozonated water. They found that hot water washing was more effective in reducing aerobic plate counts for ozone (1.30 and 1.14 log, respectively). Stivarius et al. (2002) examined the microbial ecology of ground beef and the effects of ozone versus chlorine dioxide for decontaminating beef trimmings. They also looked at the color and odor characteristics of the meat. Salmonella Typhimurium (ST) and Escherichia coli (EC) were used to inoculate beef trimmings. The trimmings were treated with 200 ppm chlorine dioxide (CLO) or 1% ozonated water for 7 or 15 min, respectively, and the results were compared to a control. After being ground, packaged, and analyzed for EC, ST, coliforms (CO), and aerobic plate counts (APC), the trimmings were displayed for 0, 1, 2, 3, and 7 days. All of the assessed bacterial species were reduced ($p < 0.05$) by the 15 min treatment with ozonated water and CLO treatments, whereas APC and ST were reduced ($p < 0.05$) by the 7O treatment.

The impact of aqueous ozone on beef slices inoculated with Clostridium perfringens (G+), Escherichia coli O157:H7 (G−), and Listeria monocytogenes (G+) was investigated by Novak and Yuan (2003). The samples were agitated for five minutes at 48 °C while being cleaned with ozonated water (3 ppm = 3 mg/L). Ozone treatment resulted in microbial count reductions of 1.28, 0.85, and 1.09 log for each injected microbe. When Castillo et al. (2003) sprayed an aqueous ozone solution (95 mg/L) over beef surfaces that had been inoculated with S. Typhimurium (G−) and E. coli O157:H7, they did not see any appreciable variations in the number of microorganisms when comparing the findings to the application of pure water. 1.46 log E. coli O157:H7 and 0.99 log aerobic bacteria were reduced on the surfaces of fresh beef by using chilled aqueous ozone (temperature = 4.6 °C–5.6 °C) at a concentration of 12 ppm, applying 90 s of spray every 30 min for 12 h. However, the treatment did not significantly reduce aerobic bacteria on the surfaces (Kalchayanand et al. 2019). The effectiveness of utilizing ozone during immersion freezing to increase the microbiological safety and lengthen the shelf life of grill drumsticks was assessed by Jindal et al. (1995). Water continuously recirculated in the chill tank, distributing ozone throughout the chill water.

Raw poultry surfaces came into contact with aqueous ozone; during immersion cooling (45 min at 0–4 °C), the initial ozone concentration in the chill water ranged between 0.44 and 0.54 ppm. The samples were then individually packaged and stored at 1–3 °C. Broiler drumsticks treated with ozone reduced aerobic plate count, coliforms, and E. Coli of more than 1.11, 0.91, and 0.90 logs, respectively. Pseudomonas aeruginosa, Gram-positive, and Gram-negative bacterial levels were lowered by 0.38, 1.11, and 1.14 logarithmic units, respectively. For as long as two days, ozonation prolonged the grill drumsticks' shelf life (which was deemed spoilt at log10 7.0 CFU/cm^2). The cool water of poultry was found to have much higher microbial reductions than the drumstick surface. Ozonated seawater was utilized to prevent Vibrio bacteria from infecting prawns (Blogoslawski et al. 1993). According to Chawla et al. (2007a, b), soaking peeled prawns in ozonated water proved more successful than treating them with a spray. Shrimp that were soaked in a solution containing three parts per million ozone for 60 s produced the most significant microbial decrease of both Pseudomonas sp. and total aerobic bacteria. Aerobic bacterial populations were shown to be lower than the initial counts in all investigated settings in a study on salmon fillets that involved 1, 2, and 3 spray passes with aqueous ozone solutions at concentrations of 1 mg/L and 1.5 mg/L. At a dosage of 1.5 mg ozone/L, three spray passes produced the most significant reduction (1.05 ± 0.18 log reduction at day 0). When three rounds of 1 mg/L ozone sprays were applied to salmon fillets infected with L. innocua, the amount of L. innocua counts was considerably ($p \leq 0.05$) reduced (1.17 ± 0.04 log reduction at day 0). They found that the number of passes beneath the spray nozzles affected the counts of microorganisms, with more passes leading to more decreases (Crowe et al. 2012). De Mendonça Silva and Gonçalves (2017) looked into how well freshwater fish were disinfected using ozonated water to eliminate bacteria. Whole and fillet Nile tilapia samples were submerged in cold water (11 °C) for 0 to 15 min, either with or without ozone (0 ppm—control) or 0.5 to 1.5 ppm. Parameters related to microbiology and physicochemistry were assessed. The most effective ozone concentration to lower the overall tilapia's microbiological contamination was 1.5 ppm (88.25% percent decrease) after 15 min of contact. The most significant reduction in the fillet treatment was observed with ozonated water at 1 and 1.5 ppm (77.2 and 79.49%, respectively).

Several studies have examined the impact of ozone in conjunction with other treatments. Ruíz-Delgado et al. (2020) discovered that ozonated water and alkaline electrolyzed water (0.68 ± 0.11 mg O_3/L) combined produced higher log reductions of E. coli (1.03 CFU/mL) in goat meat than ozonated water alone (0.53 CFU/mL). Using successive soaking and spraying techniques, Megahed et al. (2020) investigated the microbial killing capacity of an aqueous mixture of O_3 and O_3-lactic acid (O_3-LA) under various operating circumstances on chicken thighs contaminated with Salmonella. According to Stefanini Takanaca et al. (2023), mesophilic bacteria in tilapia fillets were decreased by 0.56 log CFU/g when a five ppm (5 mg/L) aqueous ozone solution and a five ppm chlorine solution (Cl + Oz) were combined. On the other hand, no difference was seen in the shelf life when compared to the control. Mesophilic counts of 1.40 log CFU/g were seen in prawn samples that were prepared by immersion in cold ozonated water (1 ppm, 10 min, 15 °C) and chlorinated water (5 ppm, 10 min, 15 °C) and subsequently packaged in the air (AIR) and a modified atmosphere (MAP). In MAP and ozone-treated samples, the first three days of storage showed the maximum efficacy in bacterial reduction (Gonçalves and Lira Santos 2018).

4.4   Effect of Ozone Treatments on Physicochemical and Sensory Properties

The physicochemical, sensory, and nutritional condition of meat and meat products may be impacted by ozone. The color of the beef samples' surfaces was the most apparent consequence of ozone. Based on sensory evaluation, ozonation can affect meat products differently. For example, in red meats, ozone can oxidize muscle tissues, deteriorate quality, change the surface color (causing unwanted discolorations), and increase rancidity in fatty tissues. The impact of ozone on physicochemical qualities is contingent upon various elements, including concentration, temperature, treatment duration, sample characteristics, and processing conditions (gaseous or aqueous ozone). Strong oxidants such as ozone and other reactive oxygen species (ROS) cause metmyoglobin to be produced via myoglobin oxidation (Bekhit et al. 2013; Khanashyam et al. 2021) and cause the meat to become discolored due to a drop in the CIE* color parameter (Mancini and Hunt 2005). When a beef carcass was treated with gaseous ozone (0.03 ppm) for nine days at 1.6 °C, shrinkage significantly accelerated, dropping from 17.8 to 7.38 (Greer and Jones 1989). According to Stivarius et al. (2002), redness (a*) marginally decreased in the shortest treatment of 7 min, while (L*) values increased in ground beef samples treated with 1% ozonated water for 7 min or 15 min. The effects of gaseous ozone exposure (10×10^{-6} kg O_3/h) at 4 °C for three days on ground Hanwoo meat were examined by Cho et al. (2014a). The parameters studied included color changes and thiobarbituric acid reactive compounds (TBARS). Ozone exposure decreased the CIE a* value of the samples throughout storage, and the TBARS values rose from 0.66 mg malonaldehyde/kg beef on day 1–0.79 mg/kg meat after three days of storage.

Regarding the TBARS values, Gimenez et al. (2021) reported that the treatment with gaseous ozone pulses, lasting between five and ten minutes each and given every thirty minutes for five hours using concentrations of 280 mg O_3/m^3 on beef, increased L* values in comparison to the control sample; however, the meat's red color did not change significantly. After ozone treatment, they also reported a final concentration of 0.7539 ± 0.0370 mg of malonaldehyde/kg meat. The research conducted by Cho et al. (2014a), Ayranci et al. (2020), and Giménez et al. (2021) revealed that the TBARS values of samples treated with ozone did not surpass 1 mg of malonaldehyde per kg of meat. This threshold is considered acceptable for displaying rancid flavor. During storage, Cho et al. (2014b) observed significant changes in L*, a*, and b* values of ozone-treated chicken breast samples (gaseous ozone at 10 mg O_3/h). These changes showed a decrease in L* and a* and an increase in b*. According to Muhlisin et al. (2016), duck and, to a lesser extent, chicken fillets that were kept for four days at 4° C under a gaseous ozone flow (10 mg O_3/m$_3$/h) with an automatic timer set to turn the machine on for fifteen minutes and off for one hundred fifty minutes showed significantly higher TBARS levels. According to Megahed et al. (2020), there was no discernible color change in chicken drumsticks treated with aqueous ozone (10 consecutive 4 min washes with water containing eight ppm ozone per wash). According to Ayranci et al. (2020), when the initial values were compared with those obtained after the 8 h treatment, the treatment of turkey breast meat with gaseous ozone (10 g/m^3) for up to 8 h at 22 °C caused significant changes in the various parameters; as a result, TBARS increased from 0.06 to 0.37 mg of malonaldehyde/kg meat and color parameters changed. L* went up to 41.97 from 34.43, while a* went down to 0.35 from 2.08. A spray of aqueous ozone (concentration 1.0 and 1.5 mg/L and 1, 2, and 3 passes under spray) did not disturb the characteristic pigmentation of salmon, according to research done by Crowe et al. (2012) on salmon fish. No significant differences in a* values were found between the controls and salmon-treated samples, suggesting that ozone did not cause bleaching of the red pigments. The pH and color of the fillets were unaffected by the ozonated water treatment in the De Mendonça Silva and Gonçalves (2017) investigation on Nile tilapia.

Nonetheless, a slight activation of the lipid oxidation pathway was noted, as shown by a rise in the TBARS value. The application of ozone lowers the characteristic odor that can occasionally be undesirable in fresh fish and bivalve mollusks, giving seafood a healthier image. It is wise to remember that ozone, in this instance, does not need to be utilized to conceal the inferior quality of the goods (Gonçalves 2009).

4.5 Effect of Ozone Treatment on the Seafood Industry to Improve Quality and Safety

It has been demonstrated that ozone can destroy a wide range of species, including viruses, bacteria, fungi, yeast, and parasites. It may also oxidize artificial materials, including detergents, herbicides, composite insecticides, and naturally occurring organic chemicals (Graham 1997; Guzel-Seydim et al. 2004). To lessen microorganisms on a variety of food products and contact surfaces, ozone has been employed in the food processing sector both as gaseous ozone and dissolved in water (Nash 2002; Kim et al. 1999; Guzel-Seydim et al. 2004; Chawla et al. 2007a, b). Ozone has been used in freezing rooms and warehouses to store food (meats, seafood, fruits, vegetables, cheeses, sausages, etc.). Reducing the bacteriological index in the storage above systems, increasing food durability (in refrigeration, freezing, or fresh storage), and getting rid of bacteria to prevent growth in meats or other foods, the development of mold, etc. are the main goals (Vaz-Velho et al. 2006).

Since molecular ozone and its breakdown products affect microbial intracellular enzymes, nucleic acids, and other cell components, it is widely recognized that they kill bacteria. In contrast to these benefits, ozone may have a pro-oxidant effect on fish constituents; nevertheless, prior research has indicated that phospholipids, membrane proteins, and polyunsaturated fatty acids (PUFAs) may be negatively impacted. This effect has not been well investigated. Ozone functions differently from chlorine due to the rapidity of oxidation and inactivation reactions. Ozone is a "general oxidizing agent," but chlorine specifically oxidizes specific enzyme systems (Campos et al. 2006).

Furthermore, the product created with ozone has a better sensory aspect and presentation and resists mold growth and putrefaction. Additionally, camera deodorization is achieved, which has a maintenance benefit. The best concentration range is between 2.5 and 3 ppm at a 1 and 3 °C temperature and a relative humidity of 90%. Higher concentrations would cause lipids to oxidize and release offensive odors (Tapp and Sopher 2002). A concentration of two to three parts per million is advised throughout freezing; however, one part per million is adequate for freezing maintenance. Intermittent ozonization is a very successful strategy. Outstanding outcomes were achieved using a concentration of 5 mg O_3.m^{-3} for two hours each day and 2–3 mg O_3.m^{-3} for the remainder of the day. This type yields a significant reduction in weight loss and an extension of the storage period. The fish lost 10% of their weight after four days at 3 °C and 65% relative humidity; in contrast, the fish lost just 4% of their weight in an ozonized environment during the same period at humidity levels between 84 and 90% (Rice et al. 1982). Ozone lowers the occasionally irritating smell of fresh fish and bivalve mollusks, adding a healthy element to these marine products. It is wise to remember that ozone needs not be utilized in this instance to conceal the poor quality and prevent economic fraud. Live fish were briefly pretreated with ozone (6 ppm) before being refrigerated and kept at zero and five degrees Celsius. Through a month of storage at 0 °C, sensory analysis revealed that ozone pretreatment increased their quality features and extended their storage life by 12 days (or 40%). According to Gelman et al. (2005), storing fish at 0 °C after ozone pretreatment seems to be a workable way to increase their shelf life, marketability, and export potential. According to Gelman et al. (2005), the ozonized water used to wash and dip fish or fish fillets demonstrated a significant decrease in microbial flora while not influencing the final product. In the catfish investigation, washing live fish and fillets in ozonated water resulted in highly statistically significant decreases in plate counts. According to specific reports, ozone gasses effectively disinfect surfaces.

By lowering spoilage microorganisms during mechanically peeled prawn processing activities, ozonated water treatment offers a chance to enhance product quality. It was discovered that soaking peeled shrimp meat in ozonated water worked better than simply misting the shrimp with it. Additionally, the more extended treatment periods and greater ozone concentrations were shown to be more successful in lowering the shrimp's levels of spoiling bacteria. According to Chawla et al. (2007a, b), the shrimp's lipid oxidation did not increase when ozonated water was applied. When ozone generation is used at the facility, the impact of ozonated water on ice quality produced by ice machines can be highly advantageous.

Additionally, ozonated water can be used as a water treatment method when ozone is not required in the plant (Tapp and Sopher 2002). According to Campos et al. (2006), a unique refrigeration system created by fusing an ozone generator with a slurry ice system allowed for improved sensory and microbiological quality preservation and suggested a considerable increase in the shelf life of seafood. Additionally, biochemical investigations verified that ozone had no discernible detrimental effect on fish quality (flatfish species). It permitted the prevention of specific pathways related to lipid hydrolysis and oxidation. Based on the findings, turbot and other flatfish species may benefit from using slurry ice and ozone for refrigerated storage (Campos et al. 2006). In freshwater aquaculture systems, ozone has long been used as a disinfectant. It is used to treat fish and eggs, sterilize the water (improving water quality), break down odorous compounds (geosmin and 2-methyllisoborneol, or 2-MIB) in natural waters, and, as of late, improve fish sensory quality and shelf life (Kim et al. 1999; Campos et al. 2006). Delivering clean fish in clean water with low microbial loads will be beneficial in preserving high-quality products. A thorough testing study conducted in 2002 showed that ozone could be added to aquaculture tanks to lessen fish slime and foam. The industry must evaluate and expand this technology (Tapp and Sopher 2002). In order to oxidize nitrite and organic material, improve overall water quality, and aid in the removal of solids across the micro screen filter, ozone was added to water in a recirculating rainbow trout (Oncorhynchus mykiss) culture system just before the culture tanks, according to Bullock et al. (1997), Summerfelt et al. (1997).

4.6   Future Perspectives

Ozonation is a cutting-edge technology that helps keep the food supply chain from contaminants without endangering the environment. Ozone has been widely utilized for a variety of purposes, including cleaning, disinfecting fruits and vegetables, treating water, eliminating odors, and sanitizing in-plant equipment. All regulatory obstacles to ozone's industrial application have been removed by the FDA and FSIS/USDA's regulatory approval of its use in food. Several influential factors have undoubtedly expedited the application of ozone. In food processing, it has served as an efficient sanitizer and disinfectant for fish, meat, and related goods. Its eco-friendliness, wide application, lack of hazardous residue generation, and particular handling requirements make it a more valuable technology for sanitation in the food business than conventional chemical sensitization techniques. The ozone concentration varies depending on the kind of treated materials, ambient conditions, microorganisms, and pollutants contained in the product. This is the most crucial factor to consider when using the ozone process. To advance ozonation as the only method of disinfection employed in the food processing sectors, research on product-based optimization of ozone treatment is needed.

References

Al-Haddad, K. S., Al-Qassemi, R. A., and Robinson, R. K. (2005). The use of gaseous ozone and gas packaging to control populations of Salmonella infantis and Pseudomonas aeruginosa on the skin of chicken portions. Food Control, 16(5), 405-410.

Antunes, P., Mourão, J., Campos, J., and Peixe, L. (2016). Salmonellosis: the role of poultry meat. Clin. Microbiol. Infect. 22, 110–121. https://doi.org/10.1016/j.cmi.2015.12.004.

Aponte, M., Anastasio, A., Marrone, R., Mercogliano, R., Peruzy, M. F., and Murru, N. (2018). Impact of gaseous ozone coupled to passive refrigeration system to maximize shelf-life and quality of four different fresh fish products. LWT-Food Sci. Technol. 93, 412–419. https://doi.org/10.1016/j.lwt.2018.03.073.

Aslam, R., Alam, M. D., and Afthab, P. P. (2020). Sanitization potential of ozone and its role in postharvest quality management of fruits and vegetables. Food Eng. Rev. 12, 48–67. https://doi.org/10.1007/s12393-019-09204-0.

Ayranci, U. G., Ozunlu, O., Ergezer, H., and Karaca, H. (2020). Effects of ozone treatment on microbiological quality and physicochemical properties of Turkey breast meat. Ozone Sci. engrg. 42 (1), 95–103. https://doi.org/10.1080/01919512.2019.1653168.

Batagoda, J. H., Hewage, S. D. A., and Meegoda, J. N. (2018). Nano-ozone bubbles for drinking water treatment. J. Environ. Eng. Sci. 14, 57–66. https://doi.org/10.1680/jenes.18.00015.

Bekhit, A. E. D. A., Hopkins, D. L., Fahri, F. T., and Ponnampalam, E. N. (2013). Oxidative processes in muscle systems and fresh meat: sources, markers, and remedies. Compr. Rev. Food Sci. Food S. 12, 565–597. https://doi.org/10.1111/1541-4337.12027.

Blogoslawski, W. J., Perez, C., and Hitchens, P. (1993). "Ozone treatment of seawater to control Vibriosis in mariculture of penaeid shrimp, Penaeus vannameii," in Proceedings of the International symposium on ozone-oxidation methods for water and wastewater treatment, Wasser Berlin, Paris, France, 26–28 April, 1993. Int Ozone Assoc. pp. I.5.1– I.5.11.

Brodowska, A. J., Nowak, A., and Śmigielski, K. (2017). Ozone in the food industry: principles of ozone treatment, mechanisms of action, and applications: an overview. Crit. Rev. Food Sci. Nutr. 58, 2176–2201. https://doi.org/10.1080/10408398.2017.1308313.

Bullock, G. L et al (1997), Ozonation of a recirculating rainbow trout culture system I. Effects on bacterial gill disease and heterotrophic bacteria, Aquaculture, 158 (1), 43–55.

Cameron, T., and Rice, R. G. (2012). "Generation and control of ozone," in Ozone in food processing. Editors C. O'Donnell, B. K. Tiwari, P. J. Cullen, and R. G. Rice 1 (WileyBlackwell), 33–54. https://doi.org/10.1002/9781118307472.ch4.

Campos, C. A. et al. (2006), Evaluation of an ozone– slurry ice combined refrigeration system for the storage of farmed turbot (Psetta maxima). Food Chemistry, 97, 223–230.

Castillo, A., McKenzie, K. S., Lucia, L. M., and Acuffi, G. R. (2003). Ozone treatment for reduction of Escherichia coli 0157:H7 and Salmonella serotype Typhimurium on beef carcass surfaces. J. Food Prot. 66 (5), 775–779. https://doi.org/10.4315/0362-028x-66.5.775.

Chawla, A., Bell, J. W., and Marlene, E. J. (2007). Optimization of ozonated water treatment of wild-caught and mechanically peeled shrimp meat. J. Aquat. Food Product. Technol. 16 (2), 41–56. https://doi.org/10.1300/J030v16n02_05.

Chawla, A.; Bell, J. W. and Marlene, E. J. (2007), Optimization of Ozonated Water Treatment of WildCaught and Mechanically Peeled Shrimp Meat, Journal of Aquatic Food Product Technology, 16 (2), 41–56.

Cho, Y., Muhlisin, M., Choi, J. H., Hahn, T. W., and Lee, S. K. (2014b). Bacterial counts and oxidative properties of chicken breast inoculated with Salmonella Typhimurium exposed to gaseous ozone. J. Food Saf. 35, 137–144. https://doi.org/10.1111/jfs.12161.

Cho, Y., Muhlisin, M., Choi, J. H., Hahn, T. W., and Lee, S. K. (2014a). Effect of gaseous ozone exposure on the bacteria counts and oxidative properties of ground hanwoo beef at refrigeration temperature. Food Sci. Anim. Resour. 34 (4), 525–532. https://doi.org/10.5851/kosfa.2014.34.4.525.

Coll Cardenas, F., Andres, S., Giannuzzi, L., and Zaritzky, N. (2011). Antimicrobial action and effects on beef quality attributes of a gaseous ozone treatment at refrigeration temperatures. Food control. 22 (8), 1442–1447. https://doi.org/10.1016/j.foodcont.2011.03.006.

Crowe, K. M., Skonberg, D., Bushway, A., and Baxter, S. (2012). Application of ozone sprays as a strategy to improve the microbial safety and quality of salmon fillets. Food control. 25 (2), 464–468. https://doi.org/10.1016/j.foodcont.2011.11.021.

Da Silva, M. V., Gibbs, P. A., and Kirby, R. M. (1998). Sensorial and microbial effects of gaseous ozone on fresh scad (Trachurus trachurus). J. Appl. Microbiol. 84, 802–810. https://doi.org/10.1046/j.1365-2672.1998.00413.x.

De Mendonça Silva, A. M., and Gonçalves, A. A. (2017). Effect of aqueous ozone on microbial and physicochemical quality of Nile tilapia processing. J. Food Process. Preserv. 41 (6), e13298. https://doi.org/10.1111/jfpp.13298.

Epelle, E. I., Macfarlane, A., Cusack, M., Burns, A., Okolie, J. A., Mackay, W., et al. (2023). Ozone application in different industries: a review of recent developments. Chem. Eng. J. 454, 140188. https://doi.org/10.1016/j.cej.2022.140188.

Farag, A. A., Saad M. Saad[1], Fahim A. Shaltout1, Hashim F. Mohammed(2023b): Organochlorine Residues in Fish in Rural Areas. Benha Journal of Applied Sciences , 8 (5): 331–336.

Fearnley, E., Raupach, J., Lagala, F., and Cameron, S. (2011). Salmonella in chicken meat, eggs and humans; Adelaide, South Australia, 2008. Int. J. Food Microbiol. 146, 219–227. https://doi.org/10.1016/j.ijfoodmicro.2011.02.004.

Gelman, A. et al. (2005), effect of ozone pretreatment on fish storage life at low temperature. J. Food Prot., 68, 778–784.

Gertzou, I. N., Karabagias, I. K., Drosos, P. E., and Riganakos, K. A. (2017). Effect of combination of ozonation and vacuum packaging on shelf life extension of fresh chicken legs during storage under refrigeration. J. Food Eng. 213, 18–26. https://doi.org/10.1016/j.jfoodeng.2017.06.026.

Gimenez, B., Graiver, N., Giannuzzi, L., and Zaritzky, N. (2021). Treatment of beef with gaseous ozone: physicochemical aspects and antimicrobial effects on heterotrophic microflora and Listeria monocytogenes. Food control. 121, 107602. https://doi.org/10.1016/j.foodcont.2020.107602.

Gonçalves, A. A. (2009). Ozone. An emerging technology for the seafood industry. Braz ArchBiol Techn 52 (6), 1527–1539. https://doi.org/10.1590/S1516-89132009000600025.

Gonçalves, A. A., and Lira Santos, T. C. (2018). Improving quality and shelf-life of whole chilled Pacific white shrimp (Litopenaeus vannamei) by ozone technology combined with modified atmosphere packaging. LWT-Food Sci. Technol. 99, 568–575. https://doi.org/10.1016/j.lwt.2018.09.083.

Graham, D. M. (1997), Use of ozone for food processing. Food Technology, 6 (51), 72–75.

Greene, A. K., Zeynep, B.G.-S., and Can, S. A. (2012). "Chemical and physical properties of ozone," in Ozone in food processing. Editors C. O'Donnell, B. K. Tiwari, P. J. Cullen, and R. G. Rice 1 (Wiley- Blackwell), 19–32. https://doi.org/10.1002/9781118307472.ch3.

Greer, G. G., and Jones, S. D. M. (1989). Effects of ozone on beef carcass shrinkage, muscle quality and bacterial spoilage. Can. Inst. Food Sci. Technol. J. 22 (2), 156–160. https://doi.org/10.1016/s0315-5463(89)70352-7.

Guzel-Seydim, Z. B. et al. (2004), Use of ozone in the food industry. Lebensm.-Wiss. u.-Technol. 37, 453–460.

Hassan, M.A and Shaltout, F. (2004). Comparative Study on Storage Stability of Beef, Chicken meat, and Fish at Chilling Temperature. Alex.J.Vet.Science, 20(21):21–30.

Ingram, M., and Haines, R. B. (2009). Inhibition of bacterial growth by pure ozone in the presence of nutrients. J. Hyg. 47, 146–158. https://doi.org/10.1017/s0022172400014406.

Jaksch, D., Margesin, R., Mikoviny, T., Skalny, J. D., Hartungen, E., Schinner, F., et al. (2004). The effect of ozone treatment on the microbial contamination of pork meat measured by detecting the emissions using PTR-MS and by enumeration of microorganisms. Int. J. Mass Spectrom. 239, 209–214. https://doi.org/10.1016/j.ijms.2004.07.018.

Jindal, V., Waldroup, A. L., Forsythe, R. H., and Miller, M. (1995). Ozone and improvement of quality and shelf life of poultry products. J. Appl. Poult. Res. 4, 239–248. https://doi.org/10.1093/japr/4.3.239.

Kalchayanand, N., Worlie, D., and Wheeler, T. (2019). A Novel aqueous ozone treatment as a spray chill intervention against Escherichia coli O157:H7 on surfaces of fresh beef. J. Food Prot. 82, 1874–1878. https://doi.org/10.4315/0362-028X.JFP-19-093.

Khadre, M. A., Yousef, A. E., and Kim, J. G. (2001). Microbiological aspects of ozone applications in food: a review. J. Food Sci. 66, 1242–1252. https://doi.org/10.1111/j.1365-2621.2001.tb15196.x.

Khanashyam, A. C., Shanker, M. A., Kothakota, A., Mahanti, N. K., and Pandiselvam, R. (2021). Ozone applications in milk and meat industry. Ozone Sci. Eng. 44, 50–65. https://doi.org/10.1080/01919512.2021.1947776.

Kim, J. G., Yousef, A. E., and Dave, S. (1999). Application of ozone for enhancing the microbiological safety and quality of foods: a review. J. Food Prot. 62 (9), 1071–1087. https://doi.org/10.4315/0362-028x-62.9.1071.

Lyu, F., Shen, K., Ding, Y., and Ma, X. (2016). Effect of pretreatment with carbon monoxide and ozone on the quality of vacuum packaged beef meats. Meat Sci. 117, 137–146. https://doi.org/10.1016/j.meatsci.2016.02.036.

Mancini, R. A., and Hunt, M. C. (2005). Current research in meat color: review. Meat Sci. 80, 43–65. https://doi.org/10.1016/j.meatsci.2008.05.028.

Manousaridis, G., Nerantzaki, A., Paleologos, E. K., Tsiotsias, A., Savvaidis, I. N., and Kontominas, M. G. (2005). Effect of ozone on microbial, chemical and sensory attributes of shucked mussels. Food Microbiol. 22, 1–9. https://doi.org/10.1016/j.fm.2004.06.003.

Megahed, A., Aldridge, B., and Lowe, J. (2020). Antimicrobial efficacy of aqueous ozone and ozone–lactic acid blend on Salmonella-contaminated chicken drumsticks using multiple sequential soaking and spraying approaches. Front. Microbiol. 11, 593911. https://doi.org/10.3389/fmicb.2020.593911.

Muhlisin, M., Utama, D. T., Lee, J. H., Choi, J. H., and Lee, S. K. (2016). Effects of gaseous ozone exposure on bacterial counts and oxidative properties in chicken and duck breast meat. Food Sci. Anim. Resour. 36 (3), 405–411. https://doi.org/10.5851/kosfa.2016.363.405.

Muthukumar, A., and Muthuchamy, M. (2013). Optimization of ozone in gaseous phase to inactivate Listeria monocytogenes on raw chicken samples. Food Res. Int. 54, 1128–1130. https://doi.org/10.1016/j.foodres.2012.12.016.

Muthukumarappan, K., Halaweish, F., and Naidu, A. S. (2000). "Ozone," in Natural food antimicrobial systems. Editor A. S. Naidu (Boca Raton, FL: CRC Press), 783–800. https://doi.org/10.1201/9780367801779.

Nash, B. (2002), Ozone effective in preserving seafood freshness, Marine Extension News, North Carolina Sea Grant, Spring.

Niveditha, A., Pandiselvam, R., Prasath, V. A., Singh, S. K., Gul, K., and Kothakota, A. (2021). Application of cold plasma and ozone technology for decontamination of Escherichia coli in foods-a review. Food control. 130, 108338. https://doi.org/10.1016/j.foodcont.2021.108338.

Novak, J. S., and Yuan, J. T. C. (2003). Viability of Clostridium perfringens, Escherichia coli, and Listeria monocytogenes surviving mild heat or aqueous ozone treatment on beef followed by heat, alkali, or salt stress. J. Food Prot. 66, 382–389. https://doi.org/10.4315/0362-028x-66.3.382.

Oner, M. E., Walker, P. N., and Demirci, A. (2011). Effect of in-package gaseous ozone treatment on shelf life of blanched potato strips during refrigerated storage. Int. J. Food Sci. Technol. 46, 406–412. https://doi.org/10.1111/j.1365-2621.2010.02503.x.

Pandiselvam, R., Prithviraj, V., Kothakota, A., and Prabha, K. (2022b). "Ozone processing of foods: methods and procedures related to process parameters," in Emerging food processing technologies. Methods and protocols in food science. Editor M. Gavahian (New York, NY: Humana). https://doi.org/10.1007/978-1-0716-2136-3_4.

Pandiselvam, R., Singh, A., Agriopoulou, S., Sachadyn-Król, M., Aslam, R., Lima, C. M. G., et al. (2022a). A comprehensive review of impacts of ozone treatment on textural properties in different food products. Trends Food Sci. Technol. 127, 74–86. https://doi.org/10.1016/j.tifs.2022.06.008.

Pandiselvam, R., Subhashini, S., Banuu Priya, E. P., Kothakota, A., Ramesh, S. V., and Shahir, S. (2018). Ozone based food preservation: a promising green technology for enhanced food safety. Ozone Sci. Eng. 41, 17–34. https://doi.org/10.1080/01919512.2018.1490636.

Pandiselvam, R., Sunoj, S., Manikantan, M. R., Kothakota, A., and Hebbar, K. B. (2017). Application and kinetics of ozone in food preservation. Ozone Sci. Eng. 39 (2), 115–126. https://doi.org/10.1080/01919512.2016.1268947.

Pirani, S. (2010) "Application of ozone in food industry," in Doctoral program in animal nutrition and food safety. Milan, Italy: Università degli Studi di Milano. PhD Thesis.

Prabha, V., Barma, D., Singh, R., and Madan, A. (2015). Ozone technology in food processing: a review. Trends Biosci. 8 (16), 4031–4047.

Premjit, Y., Sruthi, N. U., Pandiselvam, R., Priyanka, B. S., Rastogi, N. K., and Tiwari, B. K. (2014). Opportunities and challenges in the application of ozone in food processing. Emerg. Technol. Food Process., 335–358. https://doi.org/10.1016/b978-0-12-411479-1.00019-x.

Priyanka, B. S., Rastogi, K. N., and Tiwari, B. K. (2014)."Opportunities and challenges in the application of ozone in food processing," in Emerging technologies for food processing 2nd ed., Vol. 19, 335–358.

Qian, Y., Zhang, J. J., Liu, C. C., Ertbjerg, P., and Yang, S. P. (2022). Effects of gaseous ozone treatment on the quality and microbial community of salmon (Salmo salar) during cold storage. Food control. 142, 109217. https://doi.org/10.1016/j.foodcont.2022.109217.

Reagan, J. O., Acuff, G. R., Buege, D. R., Buyck, M. J., Dickson, J. S., Kastner, C. L., et al. (1996). Trimming and washing of beef carcasses as a method of improving the microbiological quality of meat. J. Food Prot. 59 (7), 751–756. https://doi.org/10.4315/0362-028X59.7.751.

Restaino, L., Frampton, E. W., Hemphill, J. B., and Palnikar, P. (1995). Efficacy of ozonated water against various food-related microorganisms. Appl. Environ. Microbiol. 61 (9), 3471–3475. https://doi.org/10.1128/aem.61.9.3471-3475.1995.

Rice, R. G.; Farquhar, J. W. and Bollyky, L. J (1982), Review of the applications of ozone for increasing storage times of perishable foods. Ozone: Science and Engineering, 4, 147–163.

Ruíz-Delgado, A., Roccamante, M. A., Malato, S., Agüera, A., and Oller, I. (2020). Olive mill wastewater reuse to enable solar photo-Fenton-like processes for the elimination of priority substances in municipal wastewater treatment plant effluents. Environmental Science and Pollution Research, 27, 38148-38154.

Saad S.M. , Shaltout, F. , Nahla A Abou Elroos, Saber B Elnahas. 2019: Antimicrobial Effect of Some Essential Oils on Some Pathogenic Bacteria in Minced Meat. J Food Sci Nutr Res. 2019; 2 (1): 012–020.

Shaltout, F.; Hanan M. Lamada , Ehsan A.M. Edris. (2020): Bacteriological examination of some ready to eat meat and chicken meals. Biomed J Sci & Tech Res., 27(1): 20461–20465.

Shaltout F., Mohammed Farouk; Hosam A.A. Ibrahim and Mostafa E.M. (2017a): Incidence of Coliform and Staphylococcus aureus in ready to eat fast foods. BENHA VETERINARY MEDICAL JOURNAL, 32(1): 13 - 17, MARCH, 2017.

Shaltout, F. , Amani M. Salem, A. H. Mahmoud, K. A(2013): Bacterial aspect of cooked meat and offal at street vendors level .Benha veterinary medical journal, 24(1): 320–328.

Shaltout, F. , Zakaria IM and Nabil ME.(2018): Incidence of Some Anaerobic Bacteria Isolated from Chicken Meat Products with Special Reference to Clostridium perfringens. Nutrition and Food Toxicology 2.5 (2018): 429–438.

Shaltout, F. , Zakaria, I.M., Nabil, M.E.(2017b): Detection and typing of Clostridium perfringens in some retail chicken meat products.BENHA VETERINARY MEDICAL JOURNAL,. 33(2):283–291.

Shaltout, F. ;Eldiasty, E. ; Salem, R. and Hassan, Asmaa (2016): Mycological quality of chicken carcasses and extending shelf – life by using preservatives at refrigerated storage. Veterinary Medical Journal -Giza (VMJG)62(3)1–7.

Shaltout, F. and Abdel Aziz ,A.M.(2004): Salmonella enterica Serovar Enteritidis in Poultry Meat and their Epidemiology .Vet.Med.J.,Giza,52(3):429–436.

Shaltout, F. and Hashim, M.F. (2002): Histamine in salted, Smoked and Canned Fish products. Benha Vet. Med.J.13 (1):1–11.

Shaltout, F.A.(2023): Viruses in Beef, Mutton, Chevon, Venison, Fish and Poultry Meat Products. Food Science & Nutrition Technology 8(4):1–10.

Shynkaryk, M. V., Pyatkovskyy, T., Mohamed, H. M., Yousef, A. E., and Sastry, S. K. (2015). Physics of fresh produce safety: role of diffusion and tissue reaction in sanitization of leafy green vegetables with liquid and gaseous ozone-based sanitizers. J. Food Prot. 78 (12), 2108–2116. https://doi.org/10.4315/0362-028X.JFP-15-290.

Steenstrup, L. D., and Floros, J. D. (2004). Inactivation of Escherichia coli O157: H7 in apple cider by ozone at various temperatures and concentrations. J. Food Process. Preserv. 28, 103–116. https://doi.org/10.1111/j.1745-4549.2004.tb00814.x.

Stefanini Takanaca, M., Corrêa Albergaria, F., Fernandes Oliveira, D. C., Mendes Ramos, E., Solis Murgas, L. D., de Sousa Gomes, M. E., et al. (2023). Microbiological and physicochemical quality of tilapia fillets treated with ozone and chlorine solution and stored under refrigeration. Food Chem. Adv. 3, 100371. https://doi.org/10.1016/j.focha.2023.100371.

Stivarius, M. R., Pohlman, F. W., Mc Elyea, K. S., and Apple, J. K. (2002). Microbial, instrumental color and sensory color and odor characteristics of ground beef produced from beef trimmings treated with ozone or chlorine dioxide. Meat Sci. 60, 299–305. https://doi.org/10.1016/s0309-1740(01)00139-5.

Summerfelt, S. T.; Hankins, J. A.; Weber, A. L. and Durant, M. D. (1997), Ozonation of a recirculating rainbow trout culture system II. Effects on microscreen filtration and water quality, Aquaculture, 158 (1), 57–67.

Tapp, C. and Sopher, C. D. (2002), Ozone Applications in Fish and Seafood Processing – Equipment Suppliers Perspective – Summary Paper. In: Ozone Applications in Fish Farming, EPRI, Palo Alto, CA: 2002, 4 pages

Vaz-Velho, M. et al. (2006), Inactivation by ozone of Listeria innocua on salmon-trout during cold-smoke processing, Food Control, 17, 609–616.

Victorin, K. (1992). Review of the genotoxicity of ozone. Mutat. Res. 277, 221–238. https://doi.org/10.1016/0165-1110(92)90045-b.

Xue, W., Macleod, J., and Blaxland, J. (2023). The use of ozone technology to control microorganism growth, enhance food safety and extend shelf life: a promising food decontamination technology. Foods 12, 814. https://doi.org/10.3390/foods12040814.